I0486421

Shortcut to Ordinary Differential Equations

Scott Imhoff

Outskirts Press, Inc.
Denver, Colorado

The opinions expressed in this manuscript are solely the opinions of the author and do not represent the opinions or thoughts of the publisher. The author has represented and warranted full ownership and/or legal right to publish all the materials in this book.

Shortcut to Ordinary Differential Equations
All Rights Reserved.
Copyright © 2009 Scott Imhoff
V2.0

This book may not be reproduced, transmitted, or stored in whole or in part by any means, including graphic, electronic, or mechanical without the express written consent of the publisher except in the case of brief quotations embodied in critical articles and reviews.

Outskirts Press, Inc.
http://www.outskirtspress.com

ISBN: 978-1-4327-1200-6

Outskirts Press and the "OP" logo are trademarks belonging to Outskirts Press, Inc.

PRINTED IN THE UNITED STATES OF AMERICA

Contents

Preface

The shortest distance between two points is a straight line (unless you are on a curved surface). This book is intended to provide the shortest path to an understanding of introductory ordinary differential equations (ODE); however, such a shortcut requires more than the shortest possible book. In recent years the understanding of how people learn has improved dramatically. Out of this enhanced understanding, *active learning* has emerged as a powerful educational tool. In preparing this book, active learning principles have been applied so that this book is a short path to a solid introductory understanding of ODE.

The active learning features of are book are as follows: The book consists of twenty short *articles*, each of which focuses on a single subject. The conceptual integrity this provides helps the student develop an organizational framework for the material. Within each Article, the material is illustrated with salient examples. Several exercises are provided at the end of each section and all of the answers are provided. This is done to provide the necessary quick reinforcement of learning and to motivate the student by early success. Thus, in keeping with active learning, the student is *actively* working with the material, reading short articles, working through short examples and exercises, obtaining immediate reinforcement, and gaining increased confidence and mastery. The book is kept short so that the student can be confident that they can "own the subject" in a reasonable span of time.

This book is intended for anyone who has had some calculus, at least to the point of being able to differentiate and integrate, and who is motivated to work the examples and exercises.

The material covered here grew out of sophomore-level classroom and hybrid courses in differential equations that the author taught at Community College of Aurora in Colorado.

The book is divided into three Parts. The first two parts treat first-order and higher-order differential equations, respectively, in the context of solution using *in situ* methods. By *in situ*, I mean that the solution is solved *in place* by some manipulation or substitution but without a major transformation of the problem into another domain. The third part of the book deals with the Laplace transform method of solution of ordinary differential equations. Here the differential equation is transformed into a realm in which many functions become rational functions and in which solution is often easier to accomplish.

Scott A. Imhoff, Ph.D.
Tewksbury, Massachusetts

Acknowledgements

Shawna Mahan of Pike's Peak Community College deserves a large measure of gratitude for her contributions to this book. She made many improvements to the manuscript, going over it with a fine-toothed comb.

I'd also like to express my thanks to Laura Young at Outskirts Press for excellent work.

Thanks especially to my wife, Sher, and our two sons, William and Frank, for their support during the many hours required to write this book.

Part one:
First-order ordinary differential equations

Article 1. Introduction: What is a differential equation?

We set about our study of differential equations by starting with an informal definition of what constitutes a differential equation. For present purposes, the following will do:

Definition 1-1: A **differential equation** is an equation involving differentials (e.g., dx, dy, etc.) in which arbitrary constants are suppressed.

An example differential equation is:

$$x\,dx + dy = 0 \qquad (1.1)$$

The above is still a differential equation if we manipulate the Leibnitz differentials in order to obtain:

$$x + \frac{dy}{dx} = 0 \qquad (1.2)$$

Or if we use Lagrange's prime notation to write:

$$x + y' = 0 \qquad (1.3)$$

We have to be careful, however, about the latter part of Definition 1-1 regarding arbitrary constants.

An example of what we mean by *arbitrary constant* in the above definition is the familiar C that arises when we evaluate the indefinite integral $\int f(x)dx$ in calculus. Differential equations are written so that these integration constants are not shown. Although derived by integrating Eq. 1-1, the following is not a differential equation since C is not suppressed:

$$-\frac{x^2}{2} + y = C \qquad (1.4)$$

Already our definition, although simplistic and informal, belies something about the heart of differential equations. Involving differentials and therefore necessarily change, they describe the dynamics of systems. Differential equations are written so as to be the most concise descriptions of general mathematical systems that undergo change. The constants of integration, the C's, which are about particular cases of systems, such as the initial stretch of a spring in a block-spring system, are set aside and the essential meaning of the equation is brought to light. Differential equations bespeak natural and mathematical truths that transcend the particulars of any given application. Differential equations represent broad, overarching generalities about systems that span many types of applications and so express profound insights about generic natural and mathematical systems.

As an example, a stretched spring is just one of many applications described by a very general differential equation known as the *harmonic oscillator*:

$$\frac{d^2 y}{dt^2} = -k\,y$$

$$(1.5)$$

This equation is simply a description of how the class of system it describes goes about restoring itself when displaced. The equation describes a system that restores itself by accelerating in negative proportion to the displacement: it pulls back without jerk.

As another example, consider the exponential decay equation:

$$y = -b\frac{dy}{dt}$$

$$(1.6)$$

What this equation describes is the change in the amount of some quantity y. Simply speaking, the rate of change in the amount (of some quantity y) is in negative proportion to how much of y is at hand (at time t). If there is a lot of material, y is decaying rapidly; if there is a small amount, y is decaying more slowly. For such a system, the amount of material y never goes to zero completely.

Both Eq. 1-5 and Eq. 1-6 are compact descriptions of important systems of vastly broad generality. The harmonic oscillator, Eq. 1-5, describes everything from capacitor tuning of an AC circuit to the vibrations of displaced atoms in a crystal lattice[1]. On the other hand, the applicability of the exponential decay model, Eq. 1-6, runs the gamut from describing the rate at which neutrons decay into protons, electrons, and anti-neutrinos[2] to predicting the rate of disappearance of froth in a mug of beer[3].

The distinction between Eq. 1-1 and Eq. 1-4 is that Eq. 1-1 is a differential equation and Eq. 1-4 is its *solution*. To see this, rearrange Eq. 1-4 so that the dependent variable is isolated on one side of the equation:

$$y = \frac{x^2}{2} + C$$

(1.7)

Thus, while solving a differential equation, one's focus changes from the general to the specific. Once we start dealing with constants of integration, we have passed from the general problem to the specifics and details of a particular case.

Details are important for applications: Being able to solve differential equations empowers you to solve real world problems in technology. Generalities are important too. Often they provide the insights into natural and mathematical systems that inspire true innovation in science. Understanding the general formalism often provides the ability to see similarities in apparently dissimilar systems and this suggests ways to transfer techniques from one problem domain to another. For example, Albert Einstein, through his deep understanding of the harmonic oscillator equation, saw how it could be applied to explain the heat capacity of solids. The first successful application of quantum mechanics to solid state physics was Einstein's Theory of Heat Capacity which was developed by treating each atom in a non-metallic solid as a harmonic oscillator[4]. Thus block-and-spring techniques were applied to a completely different level of problem, the heat capacity, by recognizing that the underlying mathematical structure for both types of systems was the harmonic oscillator.

To gain the full benefit of a course in differential equations, it is necessary to appreciate both the theory, the great generalities the equations represent, and the applications, the solutions to the equations. At the risk of oversimplifying matters, being able to solve differential equations allows one to provide basic engineering solutions. Understanding the deeper insights embodied in the equations themselves allows one to model systems and occasionally to achieve great innovation. We will try to do justice to both viewpoints in this introductory course.

---

[1] Joseph Callaway, **Quantum Theory of the Solid State**, Academic Press, New York, 1974, p.1.

[2] W. E. Burcham, **Elements of Nuclear Physics**, Longman House, Burnt Mill, U.K., 1981, p. 59.

[3] A. Leike, "Demonstration of the exponential decay law using beer froth," **European Journal of Physics**, Vol. 23, Dec. 2002.

[4] Jennifer Bothamley, **Dictionary of Theories**, Canton, MI: Visible Ink Press, 1993, p.170.

Article 2. Verifying a solution

A starting point for hands-on work with differential equations is learning to verify solutions of differential equations. A solution of a differential equation is an explicit expression for the dependent variable that makes the equation a true sentence for some specific values of the independent variables. Suppose someone shows you a differential equation and asks if some particular function $y = f(x)$ is a solution of it for some range of the independent variable x. How do you verify that $y = f(x)$ actually works? One approach is to plug $y = f(x)$ into the equation and see if the result is a true sentence for the values of the independent variable(s). Sometimes it's possible to verify that a function $y = f(x)$ is a solution to a differential equation simply by integrating the equation. Thus Eq. 1-4 is a solution to Eq. 1-1.

Example 2-1: Consider the following equation:

$$x y + y' = 0 \tag{2.1}$$

A solution, over the reals, is supposed to be:

$$y = C e^{-\frac{x^2}{2}} \tag{2.2}$$

Let's check it:

$$x y + y' = x C e^{-\frac{x^2}{2}} + C\left(-\frac{2x}{2}\right) e^{-\frac{x^2}{2}} = 0 \tag{2.3}$$

The sentence is true for all real x; therefore, Eq. 2-2 is a solution to Eq. 2-1.

◊

Example 2-2: Is $y = C x^2$ a solution to Eq. 2-1 over any interval?

Here we have:

$$y' = C 2 x \tag{2.4}$$

So, plugging into Eq. 2-1, we obtain:

$$x y + y' = x C x^2 + C 2 x \tag{2.5}$$

This expression only equals zero, making Eq. 2-1 true, only if $x = 0$. There is no finite interval over which $y = Cx^2$ is a solution. (We might add that expressing the solution as $y = Cx^2$ where x is a *variable* doesn't make sense since x cannot *vary* from 0 and the equation be true.)

EXERCISES

Exercise 2-1: Neutron decay. Given enough time, a neutron n will decay to form a proton p^+, an electron e^-, and an anti-neutrino $\overline{V}$:

$$n \rightarrow p^+ + e^- + \overline{V}$$

$$(2.6)$$

Suppose y_0 is the starting number of neutrons at time $t = 0$. The time at which only half of the original count of neutrons is present is called the *half-life* and is denoted t_{half}. Verify that:

$$y = y_0 e^{-t\frac{\ln 2}{t_{half}}}$$

$$(2.7)$$

Is a solution to the decay law for radioactive decay:

$$y + \left(\frac{t_{half}}{\ln 2}\right)\dot{y} = 0$$

$$(2.8)$$

Exercise 2-2: Harmonic oscillator. Verify that second-order equation $y'' + y = 0$ has the following two-parameter family of solutions:

$$y = C_1 \cos x + C_2 \sin x$$

$$(2.9)$$

Exercise 2-3: Special Relativity. A space capsule has length y_0 when it is at rest. As the capsule is brought to a higher and higher speed v, the length y of the capsule--as viewed by an observer at rest-- is a function of the speed v, that is, $y = y(v)$. (Here the length y is considered parallel to the direction of motion of the capsule.) Einstein's Special Theory of Relativity holds that the relationship is:

$$v + kyy' = 0 \quad \text{where} \quad k = \frac{c^2}{y_0^2}$$

$$(2.10)$$

Show that a solution to this equation is:

$$y = y_0\sqrt{1 - \frac{v^2}{c^2}}$$

(2.11)

Eq. 2-11 is known as a Lorentz Transformation[5]. The constant c is the speed of light.

---

[5] Keith R. Symon, **Mechanics**, Reading, Massachusetts: Addison-Wesly Publishing Company, 1971, p.522

Article 3. Separable variables

Differential equations grew out of the calculus and gradually became a separate subject. Isaac Newton solved some differential equations; however, it was Gottfried Leibnitz who launched differential equations as a subject *in se*. Notwithstanding Newton's staggering contributions to knowledge, it was Leibnitz who introduced the differential notation dy/dx for the derivative--a notation which was more conducive than Newton's dot notation for developing differential equations. One of Leibnitz's first methods for solving differential equations was *separable variables*; however, although Leibnitz had the seed idea for this method, it was John Bernoulli who fully developed and explained the technique.

Isaac Newton
(1643-1727)

Gottfried Leibnitz
(1646-1716)

John Bernoulli
(1667-1748)

Suppose a differential equation involves variables y and x and their differentials. To solve a differential equation using separable variables, manipulate the equation so as to place all of the terms involving y together on one side of the equals sign (or on one side of a plus or minus sign) and all of the terms involving x on the other side of the equals sign (or the plus or minus sign). Then integrate.

Example 3-1: Solve the following differential equation:

$$x\,y \; + \; y' \; = \; 0 \tag{3.1}$$

Solution: Using Leibnitz differential notation we have:

$$x\,y \; + \; \frac{dy}{dx} \; = \; 0 \tag{3.2}$$

If we multiply by dx and apply the chain rule we obtain:

11

$$x\,y\,dx \;+\; dy \;=\; 0 \tag{3.3}$$

Dividing through by y (which requires that y $\neq$ 0) we obtain:

$$x\,dx \;+\; \frac{1}{y}\,dy \;=\; 0 \tag{3.4}$$

Or, segregating the variables to either side of the equals sign we have:

$$\frac{1}{y}\,dy \;=\; -x\,dx \tag{3.5}$$

Integrating, we obtain a solution:

$$\ln(y) \;=\; -\frac{x^2}{2} \;+\; \ln C \tag{3.6}$$

Exponentiating, we obtain a solution:

$$y \;=\; C\,e^{-\frac{x^2}{2}} \tag{3.7}$$

$$\Diamond$$

Separable variables do not depend on having the variables named x and y nor does it depend on placing all of one variable on one side of the equation (We could have integrated Eq. 3-4 as is.). The separation we require is achieved when the variables, whatever their names are, are sufficiently pulled apart so that solution by integration is immediately available or one or two trivial manipulations away from being integrable.

Definition 3-1: A differential equation exhibits **separable variables** if it can be written in the form:

$$Q(y)\,dy \;+\; P(x)\,dx \;=\; 0 \tag{3.8}$$

Or:

$$\frac{dy}{dx} \;=\; g(x)\,h(y) \tag{3.9}$$

Example 3-2: Solve the differential equation:

$$y' = -y^3 e^x \tag{3.10}$$

Solution: Rewriting using Leibnitz differentials, and bringing the right-hand term over to the left we obtain:

$$\frac{dy}{dx} + y^3 e^x = 0 \tag{3.11}$$

Multiplying through by $y^{-3} dx$ and applying the chain rule, we get:

$$y^{-3} dy + e^x dx = 0 \tag{3.12}$$

This is now in the form of Eq. 3-8 and can be integrated to obtain:

$$\frac{y^{-2}}{-2} + e^x = -\frac{1}{2}C \tag{3.13}$$

This may be solved for y:

$$y = \frac{1}{\sqrt{C + 2e^x}} \tag{3.14}$$

◊

Note that in both of these examples, we chose to write constants of integration in a form convenient for our purposes: In the first example we chose *lnC* and in the second example we chose -½C. We can get away with this since *lnC* can be made to represent any real number. The same is true for -½C.

Example 3-3: Solve the nonlinear first order differential equation:

$$y' - y^2 + 9 = 0 \tag{3.15}$$

Solution: We can recast Eq. 3-15 as:

$$\frac{dy}{dx} - \left(y^2 - 9\right) = 0 \tag{3.16}$$

13

Applying dx and dividing by $(y^2 - 9)$, we have:

$$\frac{dy}{(y - 3)(y + 3)} - dx = 0 \tag{3.17}$$

Using partial fractions, this becomes:

$$\frac{\frac{1}{6} dy}{(y - 3)} - \frac{\frac{1}{6} dy}{(y + 3)} - dx = 0 \tag{3.18}$$

Which leads to:

$$\frac{1}{6} \ln(y - 3) - \frac{1}{6} \ln(y + 3) - x = \ln C \tag{3.19}$$

Or:

$$\frac{y - 3}{y + 3} = C e^{6x} \tag{3.20}$$

This becomes:

$$y = 3 \frac{1 + C e^{6x}}{1 - C e^{6x}} \tag{3.21}$$

◊

There are many types of differential equations for which separable variables will not yield a solution; however, separation of variables is useful often enough for us to add it to a collection of tools for solving differential equations. Separable variables is the first of many tools for solving differential equations that we will gather together in our knowledge base. As we go forward we will add many other tools to our kit. Equipped with any single tool from the collection, we will be only lightly armed to tackle the mainstream of ordinary differential equations; however, once we have collected a dozen or so major tools of the trade, we will be able to hold forth admirably against a variety of problem types.

EXERCISES

Exercise 3-1: Treating b as a constant, solve the exponential decay equation by separation of variables:

$$y + b\frac{dy}{dt} = 0 \tag{3.22}$$

Exercise 3-2: Solve the following:

$$\frac{dy}{dx} = y^2 - 4 \tag{3.23}$$

Exercise 3-3: Using separation of variables, solve the relativity equation in which the length of an object y is a function of its velocity v:

$$v + kyy' = 0 \quad \text{where} \quad k = \frac{c^2}{y_0^2} \tag{3.24}$$

(Use the fact that, at $v = 0$, the length is $y = y_0$ to simplify your result.)

Exercise 3-4: Solve:

$$2xy\,dx + x^2\,dy + 6x\,dx = \pi\,dy \tag{3.25}$$

Photo Credits:
http://www0.york.ac.uk/depts/maths/histstat/people/bernoulli_tree.htm

Article 4. Classifying equations and solutions

As with any mathematical endeavor, it is useful to classify the problems into different types. This classification helps guide decision making about what types of tools to use to work with the equation. There are several types of differential equations that we will need to distinguish among. There are also a few classes of equations that fall just outside the category of differential equations but which require mention since they are closely related.

This text is concerned primarily with *ordinary differential equations* which can be defined as follows:

Definition 4-1: An **ordinary differential equation (ODE)** is a differential equation in which the derivatives are with respect to one independent variable.

The following are examples of ordinary differential equations:

$$\frac{dy}{dx} + 5y = e^x \tag{4.1}$$

$$\frac{d^2y}{dx^2} - \pi\frac{dy}{dx} + 6y = 0 \tag{4.2}$$

Another class of differential equation occurs when multiple partial derivatives of a multivariate function appear. These equations, called *partial differential equations*, are usually not studied in depth until after a course in ordinary differential equations has been completed.

Definition 4-2: A differential equation involving derivatives of a multivariate function with respect to more than one of its independent variables is a **partial differential equation (PDE)**.

For a shape $y = y(x, t)$ undergoing melting, a function of position x and time t, the *heat equation* is obeyed. This is a partial differential equation:

$$\frac{\partial^2 y}{\partial x^2} = \frac{\partial y}{\partial t} \tag{4.3}$$

The primed notation for indicating the order of a derivative will often be useful. One prime means one derivative with respect to the independent variable. Two primes mean second derivative. Three primes indicate three derivatives. For four derivatives and higher, we put the number of derivatives in parentheses as a superscript of the function being differentiated:

$$y' \qquad y'' \qquad y''' \qquad y^{(4)} \qquad y^{(5)}$$

first derivative $\qquad$ second derivative $\qquad$ third derivative $\qquad$ fourth derivative $\qquad$ fifth derivative $\qquad$ (4.4)

There is another useful notation (of Newton) which involves placing a dot over a variable being differentiated with respect to time. The "fly speck" notation *should only* be used when it's assumed that the differentiation is with respect to time:

$$\dot{y} = \frac{dy}{dt}, \qquad \ddot{y} = \frac{d^2 y}{dt^2}, \qquad \dddot{y} = \frac{d^3 y}{dt^3} \qquad (4.5)$$

An important way to classify a differential equation is to identify the order of the equation.

Definition 4-3: The **order** of a differential equation is the highest derivative in the equation (the largest number of times that the dependent variable is differentiated).

The equation below is first order:

$$(y - 2x)dx + 3x\,dy = 0 \qquad (4.6)$$

The following equation is fourth order:

$$x\,y^{(4)} + 7y^2 = x^3 \qquad (4.7)$$

It's easy to be tricked when trying to decide the order of an equation. Try not to be confused between the largest number of times the dependent variable is differentiated (the order) and the largest power that a term is raised to. For example, the equation below is second order (not third order).

$$\left(\frac{d^2 y}{dx^2}\right)^3 + 5\frac{dy}{dx} = e^{-x} \qquad (4.8)$$

There are a couple of ways of writing differential equations that mathematicians find particularly decorous. In the **normal form**, the highest order derivative appears on the left of the equals sign while the lower order items are relegated to the right:

$$\frac{d^n y}{dx^n} = f\left(x, y, y', ..., y^{(n-1)}\right) \quad \text{normal form} \quad (4.9)$$

In the **general form**, all the terms are arranged as one big function F which is equal to zero:

$$F(x, y, y', ..., y^{(n)}) = 0 \quad \text{general form} \quad (4.10)$$

Another important taxonomy of differential equations involves their *linearity*. Linear differential equations are generally much more approachable than nonlinear differential equations.

Definition 4-4: A **linear differential equation** has the property that the derivative of highest order is a linear function of any lower-order derivatives that may be present.

Not surprisingly, with the linearity concept, we also have to speak about the **order of linearity**. Think of a line $z = my + b$ in the y-z plane. When we say that a differential equation is first-order linear, we are only saying that it is *formally* similar to a line such as $z = my + b$. Here dy/dx plays the role of z:

$$\text{First - order linear}: \quad \frac{dy}{dx} = m(x)y + b(x) \quad (4.11)$$

For second-order linear differential equations, the second derivative is a linear function of the lower order terms dy/dx and y.

$$\text{Second - order linear}: \quad \frac{d^2 y}{dx^2} = p(x)\frac{dy}{dx} + q(x)y + r(x)$$

$$(4.12)$$

Note that, for this generalized concept of linearity, the "coefficients" $m(x)$, $p(x)$, $q(x)$, and $r(x)$ do not have to be constants. The general n-th order linear differential equation may be written as follows:

n^{th} - order linear :

$$\frac{d^n y}{dx^n} = a_{n-1}(x)\frac{d^{n-1}y}{dx^{n-1}} + a_{n-2}(x)\frac{d^{n-2}y}{dx^{n-2}} +...+ a_0 y + a_{-1}$$

(4.13)

Example 4-1: Some examples of linear and nonlinear differential equations. Look at these examples carefully. They may look similar; however, they are drastically different equations.

(4.14)

Linear		Nonlinear	
$\frac{dy}{dx} + (\sin x)y = 0$		$\frac{dy}{dx} + (\sin y)x = 0$	
$\frac{d^2 y}{dx^2} + x^2 y = 0$		$\frac{d^2 y}{dx^2} + xy^2 = 0$	

We are now in a position to state a little more carefully what constitutes a solution of a differential equation.

Definition 4-5: The function $y = \phi(x)$ is a **solution** to the equation F(x, y, y', ..., $y^{(n)}$) = 0 on the interval I (on the x-axis) if:

    a.) $\phi(x)$ has at least n continuous derivatives.
    b.) $\phi(x)$ satisfies the equation (on the interval I).
Example 4-2: The equation

$$y + xy' = 0$$

(4.15)

Has solution:

$$y = \frac{C}{x}, \qquad x \in I = (-\infty, 0) \cup (0, \infty)$$

(4.16)

◊

Note how care has been taken to exclude the point $x = 0$ from the domain of the solution.

Many times the flat line y = 0 will solve a differential equation. (The example above being a case in point.) The solution y = 0 is known as the **trivial solution**. The trivial solution is aptly named because it can often be found by inspection. Also, the trivial solution often represents an unimportant mode of the differential equation (i.e., the block-spring system when the spring is not stretched and everything remains at rest).

In addition to characterizing differential equations, we should say a few words to characterize solutions. An **explicit solution** is one in which the dependent variable is isolated on one side of the equation. In other words y = ϕ(x).

Example 4-3:

$$y = \sqrt{1 - x^2}, \ -1 < x < 1 \tag{4.17}$$

Is an explicit solution to:

$$x + yy' = 0 \tag{4.18}$$

◊

An **implicit solution** is an equation which solves the differential equation (and does not contain differentials or derivatives) but is not in the form y = ϕ(x).

Example 4-4:

$$x^2 + y^2 = 1 \tag{4.19}$$

Is an implicit solution to:

$$x + yy' = 0 \tag{4.20}$$

◊

Different differential equations may differ in the number of degrees of freedom their solutions have. For an *n*-th order linear differential equation, we will be looking for an **n-parameter family of solutions** rather than a single solution. The first-order differential equation xy + y' = 0 has the following one-parameter family of solutions:

$$y = Ce^{-\frac{x^2}{2}} \tag{4.21}$$

The second-order equation $y'' + y = 0$ has the following two-parameter family of solutions:

$$y = C_1 \cos x + C_2 \sin x \tag{4.22}$$

A **particular solution** is one containing no parameters. The following example is a harmonic oscillator with a forcing function 2 cos x.

Example 4-5: Show that function $y = x \sin x$ is a particular solution of $y'' + y = 2\cos x$.

Solution: Differentiating the candidate solution function, we have:

$$y' = \sin x + x \cos x \tag{4.23}$$

A second differentiation yields:

$$y'' = \cos x + \cos x - x \sin x = 2\cos x - x \sin x \tag{4.24}$$

Plugging into $y'' + y$ we obtain:

$$y'' + y = 2\cos x - x \sin x + x \sin x = 2\cos x \tag{4.25}$$

Which shows that the equation $y'' + y = 2\cos x$ is satisfied.

$\Diamond$

There are some non-differential equations that warrant mentioning. **Difference equations** form the foundation for digital signal processing. An example difference equation is a one-sample delay feedback loop:

$$y[n] = x[n] - y[n-1] \tag{4.26}$$

The square brackets indicate that the independent variable n assumes integer values. In this course we will learn how to use Laplace transforms to solve higher-order differential equations. The Laplace method that we will be learning is extremely close to a method for solving difference equations called the z-transform method. (In fact the Laplace transform is a special case of the z-transform).

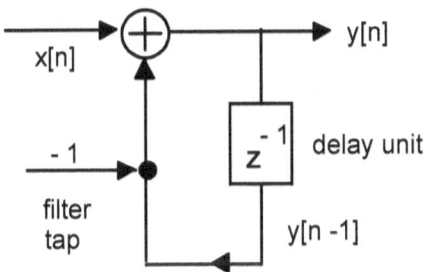

A discrete-time circuit corresponding to Eq. 4-26

Another type of non-differential equation that has come into prominence only very recently are dilation equations. They are used to generate functions called *wavelets* which are important for applications ranging from image compression to voice recognition. In dilation equations the independent variables are scaled in the argument more on one side of the equation than on the other. An example dilation equation is the following:

$$\frac{1}{\sqrt{2}}\phi(x) = h_0 \phi(2x) + h_1 \phi(2x) + h_2 \phi(2x) + h_3 \phi(2x)$$

(4.27)

Equations of this type are "solved" using recursion. The solution $\phi(x)$ is "grown" iteratively.

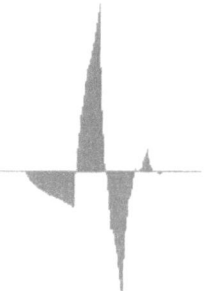

A wavelet function generated using Eq. 4-26.

Shortcut to Ordinary Differential Equations

EXERCISES

Exercise 4-1: Show that the following is a solution to Eq. 4-1 for all x:

$$y = \frac{1}{6}e^x + Ce^{-5x}$$

(4.28)

Exercise 4-2: Show that y(x, t) = e⁻ᵗ cos x is a solution to Eq. 4-3.

Exercise 4-3: Determine the orders of the following three differential equations:

$$y^2 + 1 = \dot{y}$$

(4.29)

$$(y''')^5 + x^{11}(y^{(4)})^7 = e^{-13x}$$

(4.30)

$$(x^7 y^3)^{(6)} + (x^7 y^3)^6 + (y''')^{(4)} + e^{-10x}y^{(5)}x^{19} = y^{23}$$

(4.31)

Exercise 4-4: Determine which, if any, of the following differential equations are linear differential equations. If an equation is linear, state what order linear it is.

$$y'' + y = 2\cos x$$

(4.32)

$$\dot{y} = y^2 + 1$$

(4.33)

$$\frac{d^3 y}{dx^3} - x^7 \frac{dy}{dx} = 0$$

(4.34)

$$y' = x^2 + 1$$

(4.35)

24

Article 5. Initial value problems and boundary value problems

When solving differential equations, we often can do better than to leave the solution in a form involving undetermined coefficients. When a differential equation is presented together with some additional data about the solution, the added information can often be used to provide a more specific solution. One type of bonus data that is often available is called an initial value. A particular real value of the dependent variable $y = y_0$ at some specific real value of the independent variable t_0 is an **initial value**. The initial value y_0 is a snapshot of "where the system is" at time t_0. Actually, it is correct to call $y_0 = y(t_0)$ an *initial value* even if t_0 is not the starting time of the system. Moreover, the independent variable need not be time t in order for y_0 to be an initial value. The value $y_0 = y(x_0)$ at $x = x_0$ is an initial value. A differential equation when it is presented together with an initial value of the dependent variable is an initial value problem.

Definition 5-1: An **initial value problem** (abbreviated IVP) is a differential equation and an initial value of the dependent variable (and/or a value of one or more of its derivatives) at some fixed value of the independent variable.

Example 5-1: Bell curve.
If we combine the differential equation $xy + y' = 0$ with the initial value $y(1) = 1$ we have the following initial value problem:

$$IVP: \quad xy + y' = 0 \quad \text{subject to} \quad y(1) = 1 \quad (5.1)$$

Solution: Use separable variables as in Example 3-1 to obtain:

$$y = Ce^{-\frac{x^2}{2}} \quad (5.2)$$

Plugging in the initial value $y(1) = 1$, we find that:

$$y(1) = 1 = Ce^{-\frac{(1)^2}{2}} \quad \Rightarrow \quad C = e^{\frac{1}{2}} \quad (5.3)$$

The solution then is:

$$y = e^{\frac{1}{2}(1 - x^2)} \quad (5.4)$$

◊

Note that the solution has been determined as thoroughly as possible: no parameter C appears here.

For an n-th order differential equation, we will usually require n initial values in order to completely solve the IVP. Thus for the harmonic oscillator, a second order differential equation, two initial values are needed. Often y(0) and y'(0) are given. Note that even if two IVPs have the same differential equation, they are two different problems if their initial conditions are different. The next two examples will illustrate.

Example 5-2: IVP-1; Harmonic Oscillator starting at zero position and upward velocity of 1.

Solve the following initial value problem:

$$IVP-1: \quad \ddot{y} + y = 0 \tag{5.5}$$

With initial conditions:

$$y(0) = 0 \quad \text{and} \quad y'(0) = 1 \tag{5.6}$$

A general solution to Eq. 5-5, which will be derived later, is:

$$y = C_1 \cos t + C_2 \sin t \tag{5.6}$$

Plugging in the initial values gives us two equations:

$$
\begin{array}{llllll}
y(0) = 0 & \Rightarrow & 0 = C_1 \cos(0) + 0 & \Rightarrow & C_1 = 0 \\
y'(0) = 1 & \Rightarrow & 1 = 0 + C_2 \cos(0) & \Rightarrow & C_2 = 1
\end{array} \tag{5.7}
$$

The solution of IVP-1 is then:

$$y = \sin t \tag{5.8}$$

◊

Example 5-3: IVP-2; Harmonic Oscillator starting at position y = 1 with zero velocity. Solve the following initial value problem:

$$IVP-2: \quad \ddot{y} + y = 0 \tag{5.9}$$

26

With initial conditions:

$$y(0) = 1 \quad \text{and} \quad y'(0) = 0 \qquad (5.10)$$

Again we will work with the (yet-to-be-derived) solution:

$$y = C_1 \cos t + C_2 \sin t \qquad (5.10)$$

Plugging in the initial values gives us two equations:

$$y(0) = 1 \quad \Rightarrow \quad 1 = C_1 \cos(0) + 0 \quad \Rightarrow \quad C_1 = 1$$
$$y'(0) = 0 \quad \Rightarrow \quad 0 = 0 + C_2 \cos(0) \quad \Rightarrow \quad C_2 = 0$$
$$(5.11)$$

The solution of IVP-2 is then:

$$y = \cos t \qquad (5.12)$$

◊

The solutions of these two initial value problems are plotted below. (One period is plotted.)

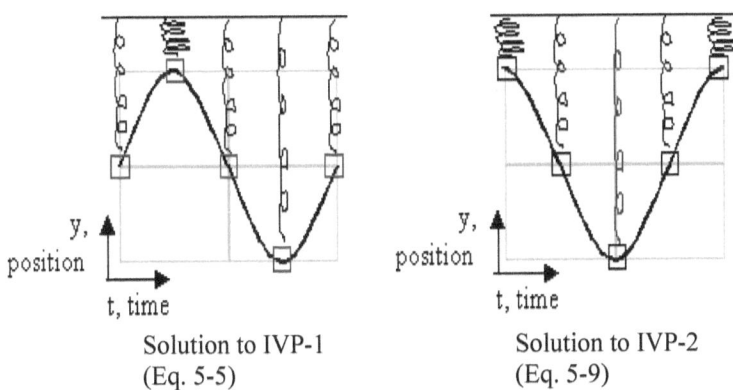

y, position → t, time	y, position → t, time
Solution to IVP-1 (Eq. 5-5)	Solution to IVP-2 (Eq. 5-9)

The harmonic oscillator equation was first recognized and solved analytically by John Bernoulli, in his work modeling a vibrating string.

In an initial value problem, we have data about the solution (its value and/or the values of one or more of its derivatives) at one fixed value of the independent variable. An equally important type of problem is the **boundary value problem**.

Shortcut to Ordinary Differential Equations

Boundary values are values of the dependent variable at two or more *different* values of the independent variable. For example we may have a differential equation and the values of the solution y at two points x_0 and x_1. Then the boundary values are $y_0 = y(x_0)$ and $y_1 = y(x_1)$.

Definition 5-2: A **boundary value problem** (abbreviated BVP) is a differential equation and two or more boundary values.

The classical example of a boundary value problem is a spring connected between two walls. At the ends where the spring is nailed down, it cannot move up or down; therefore, in a very literal sense the waves that travel on the spring are bounded. We can think outside the box though and apply the idea of boundary conditions to quite a different situation: Special Relativity.

As was mentioned in Exercise 2-2, in Albert Einstein's Special Relativity, the length of an object, y, is a function of the velocity v of the object, that is, $y = y(v)$. As the object goes to higher and higher velocities v the length y becomes shorter and shorter. The celebrated Lorentz contraction is described by the following equation:

$$ y = y_0 \sqrt{1 - \frac{v^2}{c^2}} $$

(5.14)

Where the speed of light is $c = 2.9979 \times 10^8$ m/s and y_0 is the length of the object when it is at rest ($v = 0$). As heady as Eq. 5-14 seems, we can derive it from very simple principles by recognizing a simple boundary value problem. Our starting point will be a relativity equation that simple says that the velocity times its change $v\,dv$ should be proportional (with some undetermined proportionality constant -K) to the length times its change $y\,dy$:

$$ v\,dv + K\,y\,dy = 0 $$

(5.15)

There are two boundary conditions to add to this which are conditions on the length y for different values of the independent variable v. The first boundary condition is the simple convention that the length of the object at rest is its original length:

$$ y(0) = y_0 $$

(5.16)

The second boundary condition is one of Einstein's postulates. Einstein postulated that for all observers the top speed is $v = c$. Nothing can travel faster

28

than this velocity; therefore, at the speed of light, the object *becomes* nothing. That is to say, the length at v = c is zero. This gives our second boundary condition:

$$y(c) = 0 \qquad (5.17)$$

Thus the Lorentz contraction problem is a boundary value problem. The boundary values are the lengths at the extreme values of speed: 0 and c.

Example 5-4: BVP; Special Relativity.
The boundary value problem for the Lorentz length contraction is:

$$v\, dv + K\, y\, dy = 0 \qquad (5.18)$$

Subject to:

$$y(0) = y_0 \quad and \quad y(c) = 0 \qquad (5.19)$$

Noting that Eq. 5-18 has separable variables, we integrate:

$$\frac{v^2}{2} + K\frac{y^2}{2} = \frac{C_0}{2} \quad \Rightarrow \quad v^2 + K\, y^2 = C_0 \qquad (5.20)$$

Plugging the first boundary value into Eq. 5-19, we have:

$$(0)^2 + K(y_0)^2 = C_0 \qquad (5.21)$$

The second boundary value, Einstein's postulate, when plugged into Eq. 5-20 gives us C_0:

$$(c)^2 + K(0)^2 = C_0 \quad \Rightarrow \quad C_0 = c^2 \qquad (5.22)$$

Plugging Eq. 5-22 into Eq. 5-21, we can solve for K as well:

$$K = \frac{C_0}{y_0^2} = \frac{c^2}{y_0^2} \qquad (5.23)$$

Plugging Eq. 5-22 and Eq. 5-23 back into the differential equation of Eq. 5-20 we obtain:

29

$$v^2 + \frac{c^2}{y_0^2}y^2 = c^2$$

(5.24)

Multiplying through by y_0^2 and placing the $y_0^2v^2$ term on the right, we have:

$$c^2 y^2 = y_0^2 c^2 - y_0^2 v^2$$

(5.25)

Dividing by c^2 and taking the positive square root, and factoring out y_0, we obtain:

$$y = y_0 \sqrt{1 - \frac{v^2}{c^2}}$$

(5.26)

◊

EXERCISES

Exercise 5-1: Solve the following IVP:

$$8t + 2ty + \left(t^2 - 3\right)\dot{y} = 0$$

(5.27)

With initial value:

$$y(2) = -3$$

(5.28)

Exercise 5-2: A space capsule at rest on the launch pad has a length of 10 meters. If the space capsule is brought to a speed that is 80% of the speed of light, what is the length of the space capsule? The speed of light, c, is approximately 3×10^8 meters per second.

Article 6. Existence and uniqueness of solutions

Existence and uniqueness of solutions are subjects that involve a level of rigor that tends to bog down a first course on differential equations; therefore, we will keep this article short. Our main purpose here is to suggest that existence and uniqueness cannot be taken for granted and that for some applications these issues may be vitally important.

Suppose an initial value problem describes the performance of a rocket. A rocket scientist in charge of a manned space launch solves the IVP and concludes that the rocket will take off just fine. Later, as the astronauts prepare for lift off, he becomes nervous and decides to solve the IVP again in order to check his work. This time he determines that the rocket will blow up! Unfortunately he did not make a mistake in either of his calculations: Some differential equations have two or more particular solutions. In other words, for some IVPs, the solution is not **unique**. For critical applications it is important to know about the uniqueness of the solutions of the IVP.

Like uniqueness, existence should also not be taken for granted. An example will illustrate the point.

Example 6-1:
Consider the initial value problem:

$$x y' - y \ln y = 0, \qquad y(0) = -1 \qquad (6.1)$$

Dividing through by $x \ln y$, and rearranging slightly, we get:

$$\frac{1}{\ln y} \frac{1}{y} dy = \frac{1}{x} dx \qquad \Rightarrow \qquad \ln(\ln y) = \ln x + \ln(C) \qquad (6.2)$$

Or:

$$y = e^{Cx} \qquad (6.3)$$

But this result is incompatible with y(0) = -1; therefore, there's no solution to the IVP.

Sometimes we have a solution but it's not unique:

Shortcut to Ordinary Differential Equations

Example 6-2: Consider the initial value problem:

$$y' - xy^{\frac{1}{2}} = 0 \qquad (6.4)$$

With initial condition:

$$y(0) = 0 \qquad (6.5)$$

Solution:

The particular solution $y = 0$ solves the equation as well as another particular solution:

$$y = \frac{1}{16}x^4 \qquad (6.6)$$

So there is not a unique solution.

Picard's Theorem is a tool that we will find useful to help determine whether or not solutions to initial value problems exist and are unique; however, it only applies under the right conditions.

Theorem (Picard):

Consider the following initial value problem:

$$\frac{dy}{dx} = f(x, y) \qquad (6.7)$$

With initial condition:

$$y(x_0) = y_0 \qquad (6.8)$$

Let R be the region defined as $\{ (x, y) : a \le x \le b, c \le y \le d \}$. Let (x_0, y_0) be a point in R. Let $f(x,y)$ be continuous on R and let the partial derivative of f with respect to y, that is,

$\frac{\partial f}{\partial y}$, be continuous on R. Then there exists some interval $x_0 - h < x < x_0 + h$

32

(with h positive) in R on which a unique function y(x) solves the initial value problem.

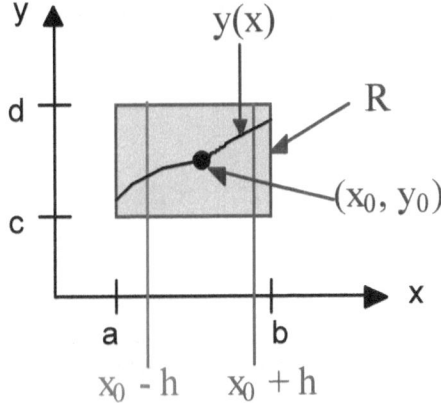

Floor plan of Picard's Theorem

We will not prove the theorem. Mostly we will limit our discussion to the method of it; however, we will say a few words to clear away some of the mystery. The theorem is about how the derivative of y, dy/dx has to be fairly well behaved in order for a solution y(x) of the differential equation to exist. We are given that a solution exists and is unique at x_0 by virtue of having the initial condition $y_0 = y(x_0)$. Picard's Theorem answers the question: What do we need to know about *f* in order to cultivate this into existence and uniqueness in a larger region.

Understanding the theorem involves a paradigm shift in our thinking. We need to think of the initial condition $y_0 = y(x_0)$ as a <u>point</u> in the x-y plane. The initial condition is the point *(x₀, y₀)*. Any solution $y = y(x)$ to the differential equation must pass through the point *(x₀, y₀)*. In the statement of Picard's Theorem, the requirements on the continuity of *f* and f_y in the region around *(x₀,y₀)* force *f* to have a certain level of good behavior around *(x₀,y₀)*. Since $dy/dx = f(x,y)$, this good behavior of *f* carries over a certain good behavior for *dy/dx* which manifests in guarantees of existence and uniqueness of the solution *y(x)*. Using the theorem involves placing the differential equation in the form of Eq. 6-6 and checking to see if the requirements are met: *f(x,y)* has to be continuous in a neighborhood R of the initial condition *(x₀, y₀)* and so does the partial derivative $f_y(x,y)$.

Example 6-1: Consider the initial value problem:

$$y' = 3y^{\frac{2}{3}} \quad \text{and} \quad y(0) = 0 \tag{6.9}$$

Cast this IVP in the form of Eq. 6-6 to see if Picard's theorem applies.

$$\text{Here} \quad (x_0, y_0) = (0, 0) \quad \text{and} \quad f(x, y) = 3y^{\frac{2}{3}} \tag{6.10}$$

Now the partial derivative with respect to y is:

$$\frac{\partial f}{\partial y} = 2y^{-\frac{1}{3}} \tag{6.11}$$

Eq. 6-11 "blows up" at (0, 0) since we have a 2/0 appearing. So Picard's theorem is not met. So we are not guaranteed existence and uniqueness. In this case it turns out that two solutions to this initial value problem are:

$$y = x^3 \quad \text{and} \quad y = 0 \tag{6.12}$$

Which underscores the failure of uniqueness for this example.

The French mathematician Émile Picard was a great connoisseur of fine food and mountain climbing (to the extent those two are compatible). Picard's mathematics resulted in significant contributions to physics, elasticity, electricity, and heat conduction.

Charles Émile Picard
(1856-1941)

Exercise 6-1: Use Picard's Theorem to see if it guarantees existence and uniqueness for the following IVP:

$$y' - xy^{\frac{1}{2}} = 0 \tag{6.13}$$

With initial condition:

$$y(0) = 0 \tag{6.14}$$

Exercise 6-2: Use Picard's Theorem to show that the following IVP has a unique solution:

$$y' + xy = 0 \tag{6.15}$$

With initial condition:

$$y(0) = 1 \tag{6.16}$$

Photo Credit:

http://es.geocities.com/fisicas/cientificos/matematicos

Article 7. Exact equations

The Englishman Isaac Newton, the German Gottfried Leibnitz, and the Swiss Bernoulli brothers founded differential equations in the late 1600's. Later, investigators in other countries began to pick up the torch, the French soon becoming enormously important. Alexis Clairaut was one of the early contributors to differential equations. We are now ready to add another tool to our box of tools for solving differential equations, one owing to Clairaut and known as *exact equations*.

Alexis Clairaut
(1713-1765)

Remember that, when you have a function of two variables, say, $f = f(x,y)$, the chain rule becomes:

$$df = \frac{df}{dx}dx + \frac{df}{dy}dy$$

$$(7.1)$$

What does this have to do with ordinary differential equations? It suggests a powerful technique for solving them known as exact equations.

If a differential equation can be written in the form:

$$M(x,y)\,dx + N(x,y)\,dy = 0 \qquad (7.2)$$

And this can also be expressed as:

$$\frac{\partial f}{\partial x}dx + \frac{\partial f}{\partial y}dy = 0$$

$$(7.3)$$

Then Eq. 7-2 is an **exact equation**. An exact equation is easy to solve. This is because, by using the chain rule, Eq. 7-3 is the same as saying $df = 0$ or $f = C$. The following theorem is a tool for being able to identify when a differential equation is exact.

Theorem (Clairaut):

Suppose we have a differential equation in the form

$$M(x,y)\,dx \quad + \quad N(x,y)\,dy \quad = \quad 0 \qquad (7.4)$$

Where M and N have continuous derivatives in x and y. And suppose also that:

$$\frac{\partial M}{\partial y} = \frac{\partial N}{\partial x} \qquad (7.5)$$

Then Eq. 7-4 is an exact equation.

Note that if we compare Eq. 7-4 and Eq. 7-3, then *exactness* means that:

$$M(x,y) = \frac{\partial f}{\partial x} \quad and \quad N(x,y) = \frac{\partial f}{\partial y} \qquad (7.6)$$

We can determine the solution of Eq. 7-2 by integrating M with respect to x and N with respect to y. We have to be careful to remember that when we integrate with respect to x, we must admit the possibility of a constant function of integration $\phi(y)$. Likewise, when we integrate with respect to y we must realize that a constant function of integration $\psi(x)$ may arise. Then we can reconcile the two integrated equations and hopefully simplify our result.

The algorithm for applying Clairaut's Theorem involves four steps: 1.) Write the equation in $Mdx + Ndy = 0$ form; 2.) Verify Clairaut's condition, i.e., $M_y = N_x$; 3.) Integrate the equations $f_x = M$ and $f_y = N$; 4.) Reconcile the expressions for f and realize $f = C$. The best way to learn is by doing.

Example 7-1: Solve the following:

$$\left(3x^2y - 2y^3 + 3\right)dx + \left(x^3 - 6xy^2 + 2y\right)dy = 0 \qquad (7.7)$$

Our first step is to see if the provisos of Theorem 7-1 are met. Looking for the $Mdx + Ndy$ form we recognize that:

$$M(x, y) = 3x^2 y - 2y^3 + 3$$
$$N(x, y) = x^3 - 6xy^2 + 2y \qquad (7.8)$$

Now the second step is to check for Clairaut's condition, namely, let's see if $M_y = N_x$:

$$\frac{\partial M}{\partial y} = 3x^2 - 6y^2 \quad \text{and} \quad \frac{\partial N}{\partial x} = 3x^2 - 6y^2 \qquad (7.9)$$

So the equation is an exact equation ... or an exact differential.

If you compare:

$$df = \frac{\partial f}{\partial x} dx + \frac{\partial f}{\partial y} dy = 0 \qquad (7.10)$$

With:

$$\left(3x^2 y - 2y^3 + 3\right)dx + \left(x^3 - 6xy^2 + 2y\right)dy = 0 \qquad (7.11)$$

You can see right away that:

$$\frac{\partial f}{\partial x} = 3x^2 y - 2y^3 + 3 \qquad (7.12)$$

And:

$$\frac{\partial f}{\partial y} = x^3 - 6xy^2 + 2y \qquad (7.13)$$

Watch how these partial derivatives are integrated. This is the third step:

$$f = x^3 y - 2xy^3 + 3x + \phi(y)$$
$$f = x^3 y - 2xy^3 + y^2 + \psi(x) \qquad (7.14)$$

The only way to reconcile these two is to have:

$$\phi(y) = y^2 \qquad \text{and} \qquad \psi(x) = 3x \tag{7.15}$$

So the solution to the differential equation is:

$$f = x^3 y - 2xy^3 + 3x + y^2 = C \tag{7.16}$$

Or is it? It is important to stop and realize that the original equation we were trying to solve was $M\,dx + N\,dy = 0$ which has nothing to do with f. f is just a helper function; therefore, the correct way to present the final solution is to only report the equation on the right in Eq. 7-15. This completes the fourth step:

$$x^3 y - 2xy^3 + 3x + y^2 = C \tag{7.17}$$

$$\lozenge$$

Example 7-2: Solve the following IVP:

$$2x\sin y - y\sin x + \left(x^2 \cos y + \cos x\right)y' = 0 \tag{7.18}$$

$$\text{subject to} \quad \frac{\pi}{6} = y\left(\frac{\pi}{2}\right) \tag{7.19}$$

Solution: First let's see if the differential equation is exact:

$$M(x, y) = 2x\sin y - y\sin x$$
$$N(x, y) = x^2 \cos y + \cos x \tag{7.20}$$

Checking Clairaut's conditions by determining M_y and N_x:

$$\frac{\partial M}{\partial y} = 2x\cos y - \sin x$$

$$\frac{\partial N}{\partial x} = 2x\cos y - \sin x \tag{7.21}$$

We see that Clairaut's condition $M_y = N_x$ is satisfied. Thus the equation $M\,dx + N\,dy = 0$ is an *exact differential* and can be expressed as $df = f_x\,dx + f_y\,dy = 0$. This permits us to solve for $f = C$ by integrating. We integrate $f_x = M$ and $f_y = N$:

$$\frac{\partial f}{\partial x} = 2x\sin y - y\sin x$$

$$\frac{\partial f}{\partial y} = x^2 \cos y + \cos x$$

(7.22)

These become:

$$f = x^2 \sin y + y\cos x + \phi(y)$$
$$f = x^2 \sin y + y\cos x + \psi(x)$$

(7.23)

Which can be reconciled by letting $\phi(y) = \psi(x) = 0$. The solution to the differential equation then is, remembering that $f = C$:

$$C = x^2 \sin y + y\cos x$$

(7.24)

But we have not yet solved the IVP. Our last step is to make use of the initial condition $\pi/6 = y(\pi/2)$:

$$C = \left(\frac{\pi}{2}\right)^2 \sin\left(\frac{\pi}{6}\right) + \frac{\pi}{6}\cos\frac{\pi}{2} \quad \Rightarrow \quad C = \frac{\pi^2}{8}$$

(7.25)

So the final solution is:

$$x^2 \sin y + y\cos x = \frac{\pi^2}{8}$$

(7.26)

_____

EXERCISES

Exercise 7-1: Use the exact equations method to solve the following:

$$x\,dx + dy = 0$$

(7.27)

Exercise 7-2: Solve the following:

$$3x^2 y\,dx + (x^3 - y^2)\,dy = 0$$

(7.28)

41

Exercise 7-3: Solve:

$$\left(2y + t\cos y\right)\dot{y} = -\sin y \qquad (7.29)$$

Exercise 7-4: Solve the IVP:

$$y\sec^2 x \, dx + \tan x \, dy = 0 \qquad (7.30)$$

With initial condition:

$$y\left(\frac{\pi}{4}\right) = 1 \qquad (7.31)$$

Exercise 7-5: Solve the following:

$$y\ln y \, dx + x \, dy = 0 \qquad (7.32)$$

Photo credit:

http://www.visitvoltaire.com/e_alexis-claude_clairaut.htm

Article 8. Linear differential equations of the first order (method of)

We have already learned two tools for solving differential equations, *separable variables* and *exact equations*; we are now ready to learn a third tool. Aptly named, the *method of first order linear equations* applies to the solution of differential equations that are first order linear. In Article 4 we said that a first order ordinary differential equation was one that can be written in the form:

$$\frac{dy}{dx} = m(x)\,y + b(x)$$

$$(8.1)$$

And so the 1st order derivative is a linear function of the 0th order derivative. Letting $m(x) = -P(x)$ and $b(x) = Q(x)$, an equivalent form for the linear equation is:

$$\frac{dy}{dx} + P(x)\,y = Q(x)$$

$$(8.2)$$

(The method of solution we are about to learn requires the form of Eq. 8-2 rather than the Eq. 8-1 form.)

The method of solution involves making the left-hand side of Eq. 8-2 "look like someone has just applied the product rule." To accomplish this state of affairs, multiply Eq. 8-2 by a certain "integrating factor":

$$e^{\int^{x} P(\psi)\,d\psi}$$

$$(8.3)$$

Multiplying Eq. 8-2 by the integrating factor, we have:

$$\frac{dy}{dx}\,e^{\int^{x} P(\psi)\,d\psi} + y\,e^{\int^{x} P(\psi)\,d\psi}\,P(x) = Q(x)\,e^{\int^{x} P(\psi)\,d\psi}$$

$$(8.4)$$

But the left hand side is exactly what you obtain when you apply the product

rule to $y\,e^{\int^{x} P(\psi)\,d\psi}$:

$$\frac{d}{dx}\left(y\, e^{\int\limits^{x} P(\psi)d\psi} \right) = \frac{dy}{dx}\, e^{\int\limits^{x} P(\psi)d\psi} + y\, e^{\int\limits^{x} P(\psi)d\psi} P(x)$$

(8.5)

Thus Eq. 8-5 can be written:

$$\frac{d}{dx}\left(y\, e^{\int\limits^{x} P(\psi)d\psi} \right) = Q(x)\, e^{\int\limits^{x} P(\psi)d\psi}$$

(8.6)

Integrating, we see that:

$$y\, e^{\int\limits^{x} P(\psi)d\psi} = \int Q(x)\, e^{\int\limits^{x} P(\psi)d\psi}\, dx$$

(8.7)

Then dividing by the integrating factor we can solve for y:

$$y = \frac{1}{e^{\int\limits^{x} P(\psi)d\psi}} \int Q(x)\, e^{\int\limits^{x} P(\psi)d\psi}\, dx$$

(8.8)

Example 8-1: Solve:

$$x^2 y' + 5xy + 3x^5 = 0$$

(8.9)

And assume that x does not equal zero. Dividing through by x^2 we have:

$$y' + 5\frac{y}{x} = -3x^3$$

(8.10)

This puts it in the proper form so that we can see that:

$$P(x) = \frac{5}{x} \qquad and \qquad Q(x) = -3x^3$$

(8.11)

So the integrating factor becomes:

$$e^{\int_{}^{x} P(\psi)d\psi} = e^{\int_{}^{x} \frac{5}{\psi}d\psi} = e^{5\ln x} = x^5 \tag{8.12}$$

Thus the solution is:

$$y = \frac{1}{x^5}\int (-3x^3)x^5\,dx = -\frac{1}{3}x^4 + \frac{C}{x^5} \tag{8.13}$$

◊

Eq. 8-8 probably makes the method look harder than it really is. To use the method it is sufficient to remember to: 1.) Place the equation in the $dy/dx + P(x)y = Q(x)$ form; 2.) Multiply through by the integrating factor $e^{\int P}$; and 3.) Reorganize the right hand side to be the derivative of a product.

Example 8-2: Solve:

$$x\frac{dy}{dx} - 7y = x^8 e^x \tag{8.14}$$

Divide through by x to get it into the $dy/dx + P(x)y = Q(x)$ form:

$$\frac{dy}{dx} - \frac{7}{x}y = x^7 e^x \tag{8.15}$$

We recognize that $P(x) = -7/x$. The integrating factor then is:

$$e^{\int_{}^{x} P(\psi)d\psi} = e^{\int_{}^{x} \frac{-7}{\psi}d\psi} = e^{-7\ln x} = x^{-7} \tag{8.16}$$

Multiplying through by x $^{-7}$, we have:

$$x^{-7}\frac{dy}{dx} - 7x^{-8}y = e^x \tag{8.17}$$

The left side of this equation is obviously the derivative of $x^{-7}y$.

45

$$\frac{d}{dx}(x^{-7}y) \;=\; e^x$$

(8.18)

Integrating we behold:

$$x^{-7}y \;=\; e^x + C$$

(8.19)

Now we can solve for y and we are done!

$$y \;=\; x^7(e^x + C)$$

(8.20)
◊

EXERCISES

Exercise 8-1: Solve:

$$\frac{dy}{dx} + 2y \;=\; e^{2x}$$

(8.21)

Exercise 8-2: Solve:

$$\cos^2 x \, \frac{dy}{dx} + y \;=\; e^{-\tan x}\cos^2 x$$

(8.22)

Exercise 8-3: Use the method of first order linear equations to solve:

$$y\,dx + dy - (\sin x)\,dx \;=\; 0$$

(8.23)

Exercise 8-4: Solve the following:

$$\frac{dy}{dx} + 2xy \;=\; e^{-x^2}$$

(8.24)

Exercise 8-5: An RL circuit initial value problem. In an electric circuit, a constant electromotive force E is provided by a battery. The circuit also contains a resistor R and an inductor L. The current in the circuit as a function of time is $j = j(t)$. The voltage drop across the resistor is jR and the voltage drop across the inductor is $j\,dL/dt$. According to Kirchoff's Law, the sum of these two voltage drops must equal the voltage supplied by the electromotive force E:

$$L\frac{dj}{dt} + R\,j = E$$

(8.25)

The current at time zero is zero, i.e., $j(0) = 0$. Use an integrating factor to solve for the current j as a function of time t: $j = j(t)$.

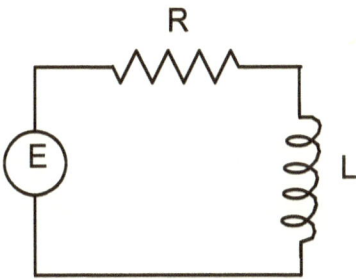

An RL circuit

Article 9. Homogeneous coefficients (method of)

Leibniz devised a method for solving equations in the form $M(x,y)dx + N(x,y)dy = 0$ that have a special property for their functions $M(x,y)$, $N(x,y)$ known as homogeneity.

Definition 9-1: A function $f(x, y)$ is **homogeneous of degree** α if, when you replace x with τx and y with τy, you obtain:

$$f(\tau x, \tau y) = \tau^{\alpha} f(x, y) \tag{9.1}$$

In other words, the τ "comes through" with power α. Think of homogeneous milk and how nicely it flows through a straw. Multipliers which scale the variables "pass through" a function in a comparable way.

Example 9-1:

$$f(x, y) = x^3 + y^3 \tag{9.2}$$

Is homogeneous of degree three because:

$$f(tx, ty) = (tx)^3 + (ty)^3 = t^3(x^3 + y^3) = t^3 f(x, y) \tag{9.3}$$

◊

Now let's talk about differential equations having *homogeneous coefficients.* You may not be accustomed to thinking about functions as coefficients; however, $M(x,y)$ and $N(x,y)$ are considered to be coefficient functions.

Definition 9-2: A differential equation of the form:

$$M(x, y)dx + N(x, y)dy = 0 \tag{9.4}$$

Is said to have **homogeneous coefficients** if M and N are homogeneous functions of the same degree.

In order to solve Eq. 9-2, we make a substitution:

$$y = ux \tag{9.5}$$

Where u is some function of x. This substitution reduces the above equation to a separable equation. We have the following:

Step 1.) Test for homogeneity (M and N should be homogeneous of the same degree.)
Step 2.) Substitute $y = u\,x$.
Step 3.) Solve the separable equation.
Step 4.) Back-substitute to eliminate u (since u is only a helping function).

Let's run this algorithm through its paces.

Example 9-2: Solve:

$$(x^2 - xy)\,dx + x^2\,dy = 0 \tag{9.6}$$

First we test for homogeneity:

$$M(tx,ty) = (ty)^2 - (tx)(ty) = t^2(y^2 - xy) = t^2 M(x,y)$$
$$N(tx,ty) = (tx)^2 = t^2 x^2 = t^2 N(x,y)$$

$$\tag{9.7}$$

So both M and N are homogeneous of degree two. The second step is to Substitute $y = ux$ (This will make the differential equation separable.):

$$(u^2 x^2 - xux)\,dx + x^2(u\,dx + x\,du) = 0 \tag{9.8}$$

Which can be written:

$$u^2 x^2\,dx - x^2 u\,dx + x^2 u\,dx + x^3\,du = 0 \tag{9.9}$$

I can cancel out the x^2 :

$$u^2\,dx - u\,dx + u\,dx + x\,du = 0 \tag{9.10}$$

The result can now be solved by separable variables:

$$u^2\,dx + x\,du = 0 \tag{9.11}$$

$$x^{-1}\,dx = -u^{-2}\,du \tag{9.12}$$

$$\ln|x| = \frac{1}{u} + \ln C \qquad (9.13)$$

The variable u was only introduced to help bridge our way to a solution; therefore, we need to substitute what it represents in order to eliminate it from our result. Thus we substitute u = y/x:

$$\ln|x| = \frac{x}{y} + \ln C \qquad (9.14)$$

We can simplify this and present our final answer:

$$x = Ce^{\frac{x}{y}} \quad \rightarrow \quad y = \frac{x}{\ln x + C} \qquad (9.15)$$

$\lozenge$

It's very straightforward to prove that the algorithm should work.

Theorem: First order homogeneous coefficient equations are reducible to separable

If $M(x,y)dx + N(x,y)dy = 0$ then, if $M(x,y)$ and $N(x,y)$ are homogeneous of the same degree, the equation can be made separable by the substitution $y = ux$.

Proof: Make the substitution $y = ux$:

$$M(x,u\,x)\,dx + N(x,ux)(x\,du + u\,dx) = 0 \qquad (9.16)$$

Then the homogeneity means that x^{α} will factor out:

$$x^{\alpha}M(1,u)\,dx + x^{\alpha}N(1,u)(x\,du + u\,dx) = 0 \qquad (9.17)$$

Dividing by x^{α}, the equation is then:

$$M(1,u)\,dx + N(1,u)x\,du + N(1,u)u\,dx = 0 \qquad (9.18)$$

Or:

$$(M(1,u) + u N(1,u)) dx + N(1,u)x \, du = 0 \qquad (9.19)$$

Dividing by N(1, u), we have:

$$(\frac{M(1,u)}{N(1,u)} + u) dx + x \, du = 0 \qquad (9.20)$$

Dividing through by these coefficient functions, we achieve separation of variables:

$$\frac{1}{x} dx + \frac{N(1,u)}{M(1,u) + u N(1,u)} du = 0 \qquad (9.21)$$

Q.E.D.

Sometimes it takes effort to overcome changes in notation; however, the methods we are learning do not depend on any particular variable names. In the following the dependent variable is x and the independent variable is t.

Example 9-3: Solve the IVP:

$$\dot{x} + 1 + \frac{9}{t}x = 0$$

With initial condition: x(10) = 1023 $\qquad (9.16)$

Here we are looking for a *M(t, x)dt + N(t, x)dx = 0* form. If we multiply through by *t* and *dt*, we're one step closer:

$$t \, dx + (t + 9x) dt = 0 \qquad (9.17)$$

We need only switch the order of the two terms and we're in *M(t, x)dt + N(t, x)dx = 0* form.

$$(t + 9x) dt + t \, dx = 0 \qquad (9.18)$$

Now we see that the coefficient functions are homogeneous of degree one:

$$M(\tau x, \tau t) = (\tau t + 9\tau x) = \tau(t + 9x) = \tau M(x, t)$$
$$N(\tau x, \tau t) = \tau t = \tau N(x, t) \qquad (9.19)$$

We have to be careful about the present change in notation. The substitution we need to make is $x = u\,t$. This is because Eq. 9-5 really tells us to replace the dependent variable with u times the independent variable. In the present problem, the dependent variable happens to be x and the dependent variable t. Substituting we obtain $x = u\,t$:

$$(t + 9ut)\,dt + t\left(u\,dt + t\,du\right) = 0 \tag{9-20}$$

We can cancel a t out of every term:

$$(1 + 9u)\,dt + u\,dt + t\,du = 0 \tag{9.21}$$

Next we can use separable variables:

$$(1 + 10u)\,dt + t\,du = 0 \tag{9.22}$$

This becomes:

$$\frac{1}{t}\,dt + \frac{1}{1 + 10u}\,du = 0 \tag{9.23}$$

Integrating:

$$\ln t + \frac{1}{10}\ln(1 + 10u) = \ln C \tag{9.24}$$

Or:

$$t(1 + 10u)^{1/10} = C \tag{9.25}$$

We now take out the *helper function* $u = x/t$:

$$t\left(1 + 10\frac{x}{t}\right)^{1/10} = C \tag{9.26}$$

The initial condition says that when $t = 10$, we have $x = 1023$. This lets us solve for C:

$$10(1 + 10 \frac{1023}{10})^{1/10} \;=\; C \;\Rightarrow\; C \;=\; 10(1024)^{1/10} \;=\; 20$$

(9.27)

The solution then is:

$$x \;=\; 2\left(\frac{20}{t}\right)^9 \;-\; \frac{t}{10}$$

$$x \;=\; 2\left(\frac{20}{t}\right)^9 \;-\; \frac{t}{10}$$

(9.28)
◊

EXERCISES

Exercise 9-1: Solve the following:

$$(x^2 + y^2)\, dx \;+\; 2xy\, dy \;=\; 0$$

(9.29)

Exercise9-2: Solve the following:

$$\left(x\cos\frac{y}{x} - y\right) dx \;+\; x\, dy \;=\; 0$$

(9.30)

Exercise 9-3: Solve:

$$x - \left(t + x\tan\frac{t}{x}\right)\dot{x} \;=\; 0$$

(9.31)

Exercise 9-4: Solve:

$$(x + 3y)\, dx \;+\; x\, dy \;=\; 0$$

(9.32)

Article 10. Bernoulli's equation

Bernoulli's equation has the basic form:

$$\frac{dy}{dx} + \psi(x)y = \phi(x)y^n \tag{10.1}$$

This can be solved by making the substitution:

$$u = y^{1-n} \tag{10.2}$$

This substitution then reduces the differential equation to a separable form. Sometimes it's easier to solve the problem by multiplying through by $(1 - n)y^{-n}$ first before making the substitution. This makes it easy to recognize u and its derivative and to substitute in order to eliminate y. Multiplying Eq. 10-1 by $(1 - n)y^{-n}$, we obtain:

$$(1-n)y^{-n}y' + (1-n)\psi(x)y^{1-n} = (1-n)\phi(x) \tag{10.3}$$

Now making the substitutions $(1-n)y^{-n}y' = u'$ and $y^{1-n} = u$, we have:

$$u' + (1-n)\psi(x)u = (1-n)\phi(x) \tag{10.4}$$

The equation is now first order linear and can be solved using an integrating factor:

$$e^{\int\limits_{}^{x} (1 - n)\psi(\xi)d\xi} \tag{10.5}$$

Example 10-1: Solve the following:

$$y' + (\tan x)y = \frac{1}{\cos^3 x}y^5 \tag{10.6}$$

In this case $n = 5$. Multiplying through by $(1 - 5)y^{-5}$, we have:

$$-4y^{-5}y' - 4(\tan x)y^{-4} = \frac{-4}{\cos^3 x} \tag{10.7}$$

55

We make the substitution $u = y^{1-5} = y^{-4}$:

$$u' - 4(\tan x)u = \frac{-4}{\cos^3 x} \tag{10.8}$$

This is now first order linear. We may determine the integrating factor as follows:

$$e^{-4\int^x \tan \xi \, d\xi} = e^{4\int^x \frac{1}{\cos \xi}(-\sin \xi) \, d\xi} = e^{4\ln(\cos x)} = \cos^4 x \tag{10.9}$$

Applying this integrating factor, we observe:

$$\left(\cos^4 x\right)u' + 4\cos^3 x\left(-\sin x\right)u = -4\cos x \tag{10.10}$$

Gathering together the left side into the derivative of a product, we have:

$$\frac{d}{dx}\left(u \cos^4 x\right) = -4\cos x \tag{10.11}$$

This is straightforward to integrate:

$$u \cos^4 x = -4\sin x + C \tag{10.12}$$

Substituting to eliminate u and solving for y, we have:

$$y = \frac{\cos x}{\left[C - 4\sin x\right]^{\frac{1}{4}}} \tag{10.13}$$

$\Diamond$

EXERCISES

Exercise 10-1: Solve:

$$x y' + 2y = 4x^4 y^4 \tag{10.14}$$

Exercise 10-2: Solve:

$$x y^3 y' + y^4 = \cos(x^4) \qquad (10.15)$$

Exercise 10-3: Solve:

$$\frac{dy}{dx} + \frac{1}{x}y = y^3 \cos\left(\frac{1}{x}\right) \qquad (10.16)$$

Part two:
Higher-order ordinary differential equations

Article 11. Euler's formula

The Swiss mathematician Leonhard Euler was possibly the greatest mathematician who ever lived. He generated about a thousand pages of new mathematics per year, including some of math's most magnificent discoveries. During the last 17 years of his life, he was totally blind; however, his output actually increased[6]. He had reached a level where he no longer needed to see; he could keep track of vast blackboards of mathematics in his mind.

Leonhard Euler
(1707-1783)

The use of the symbol e for the inverse-log of one, the use of π to express the ratio of a circle's circumference to its diagonal, and the use of i to represent the square root of negative one, are all notations introduced by Euler. Euler's formula expresses the complex exponential sinusoid $e^{i\theta}$ in terms of a real summand $cos\theta$ and an imaginary summand $i\,sin\theta$.

Euler's formula:

$$e^{i\theta} = \cos\theta + i\sin\theta$$

$$where \ \ i = \sqrt{-1} \ \ and \ \ e = 2.718281828...$$

(11.1)

Numbers in the form $a + i\,b$, where a and b are real, are called **complex numbers**. Real numbers have the property that b = 0. Pure **imaginary numbers** have the property that $a = 0$. If a, b, c, and d are real numbers, we can define complex numbers z and w as follows:

$$z = a + ib$$
$$w = c + id$$

(11.2)

To add two complex numbers, we proceed as we are used to, only being careful to keep the real terms together and the imaginary terms together:

$$z + w = a + c + i(b + d)$$
(11.3)

To multiply complex numbers we proceed using the approach of multiplying two sums, being careful to substitute $i^2 = -1$ before presenting the result:

$$zw = (a + ib)(c + id)$$
$$zw = ac + i^2 bd + iad + ibc$$
$$zw = ac - bd + i(ad + bc)$$
(11.4)

Multiplying a complex number by a scalar k has the scaling effect we expect:

$$kz = k(a + ib) = ka + ikb$$
(11.5)

Multiplying a complex number by an imaginary number ik works as follows:
$$ikz = ik(a + ib) = ika + i^2 kb = -kb + ika$$
(11.6)

The **conjugate** of a complex number z, denoted z^*, is formed by replacing i with $-i$ everywhere it occurs.

$$(a + ib)^* = a - ib$$
(11.7)

If a function $f(x, y)$ involves I, then $f(x, y)$ is a **complex function**. The **conjugate of a complex function** $f^*(x,y)$ is formed by replacing every I with $-I$ everywhere it occurs. Let's conjugate a complex exponential to illustrate:

$$h(t) = e^{i\frac{2\pi}{1024}t}$$
(11.8)

$$h^*(t) = e^{-i\frac{2\pi}{1024}t}$$
(11.9)

The **norm of a complex number** z, denoted $\|z\|$ is equal to the square root of z times its complex conjugate:

$$\|z\|^2 = z^* z$$
(11.10)

Euler's formula can be used to prove all sorts of interesting facts about complex numbers. For instance it is possible to raise the square root of negative one to the square root of negative one power using Euler's formula. The result is a real number:

Example 11-1: Reduce i^i to a real number.

Plug $\theta = \pi/2$ into Euler's formula:

$$e^{i\frac{\pi}{2}} = \cos\frac{\pi}{2} + i\sin\frac{\pi}{2} = i \tag{11.11}$$

This gives an expression for i:

$$i = e^{i\frac{\pi}{2}} \tag{11.12}$$

Raise each side of this equation to the i power:

$$i^i = \left(e^{i\frac{\pi}{2}}\right)^i = e^{i^2\frac{\pi}{2}} = e^{-\frac{\pi}{2}} \approx .21 \tag{11.13}$$

So i raised to the i is about one fifth. The solution is not unique. It's easy to prove Euler's theorem if one makes note of the fact that the powers of i follow the pattern:

$$\begin{aligned}
i^0 &= 1 \\
i^1 &= i \\
i^2 &= -1 \\
i^3 &= -i \\
i^4 &= 1 \\
i^5 &= i \\
i^6 &= -1 \\
i^7 &= -i \\
&\vdots
\end{aligned} \tag{11.14}$$

The Taylor expansions for $\cos(\theta)$ and $\sin(\theta)$ are:

$$\cos(i\theta) = 1 - \frac{(\theta)^2}{2!} + \frac{(\theta)^4}{4!} - \ldots$$

$$\sin(i\theta) = \theta - \frac{(\theta)^3}{3!} + \frac{(\theta)^5}{5!} - \ldots$$

$$(11.15)$$

If we multiply $sin(i\theta)$ by i and add it to $cos(i\theta)$, we have, interleaving the terms:

$$\cos(i\theta) + i\sin(i\theta) =$$

$$1 + i\theta - \frac{(\theta)^2}{2!} - \frac{i(\theta)^3}{3!} + \frac{(\theta)^4}{4!} + \frac{i(\theta)^5}{5!} \ldots \qquad (11.16)$$

Now using the identities in Eq. 11-12 we have:

$$\cos(i\theta) + i\sin(i\theta) =$$

$$1 + i\theta + \frac{(i\theta)^2}{2!} + \frac{(i\theta)^3}{3!} + \frac{(i\theta)^4}{4!} + \frac{(i\theta)^5}{5!} \ldots \qquad (11.17)$$

But this is the Taylor expansion for $e^{i\theta}$ which proves Euler's formula:

$$e^{i\theta} = \cos\theta + i\sin\theta$$

$$(11.18)$$

◊

Example 11-2: Determine the cube roots of negative one.

Plugging $\theta = \pi$ into Euler's formula we have:

$$e^{i\pi} = \cos\pi + i\sin\pi = -1$$

$$(11.19)$$

But since $\cos(\ )$ and $\sin(\ )$ are both periodic with period 2π, the same is true for $e^{i(\ )}$. This means we can always add $2\pi n$ to the argument without impunity:

$$-1 = e^{i(\pi + 2\pi n)}$$

$$(11.20)$$

To find the cube roots, we raise each side to the one-third:

$$(-1)^{1/3} = \left(e^{i(\pi + 2\pi n)}\right)^{1/3} = e^{i\left(\frac{\pi}{3} + \frac{2}{3}\pi n\right)}$$

(11.21)

As n assumes integer values 0, 1, 2, ..., we see that there are three cube roots and then the pattern repeats. The three distinct cube roots are:

$$(-1)^{1/3} = \begin{cases} e^{i\left(\frac{\pi}{3}\right)} \\ e^{i(\pi)} \\ e^{i\left(\frac{5}{3}\pi\right)} \end{cases}$$

(11.22)

These can be reduced using Euler's formula:

$$(-1)^{1/3} = \begin{cases} \dfrac{1}{2} + i\dfrac{\sqrt{3}}{2} \\ -1 \\ \dfrac{1}{2} - i\dfrac{\sqrt{3}}{2} \end{cases}$$

(11.23)

◊

The **inner product** between two functions g(t) and h(t) having the same domain $-L \le t \le L$, is formed by multiplying the first function times the conjugate of the second and integrating the result:

$$g(t) \cdot h(t) = \int_{-L}^{L} g(t) h^*(t) \, dt$$

(11.24)

Functions that have a zero inner product are **orthogonal functions**.

Example 11-3: The two functions sin(t) and cos(t) are orthogonal functions, since their inner product is zero:

$$\int_{-L}^{L} \cos(t)\sin(t) \, dt = -\frac{\cos^2(t)}{2}\bigg|_{-L}^{L} = 0$$

(11.25)

◊

EXERCISES

Exercise 11-1: Calculate the norm of 3 + 4i:

Exercise 11-2: Calculate the following:

$$\cos\left(\frac{\ln i}{i}\right)$$

(11.26)

Exercise 11-3: Determine the fourth roots of one.

Exercise 11-4: What is the product of the right hand side of Eq. 11-8 and the right hand side of Eq. 11-9?

Exercise 11-5: The *norm of a function* is the square root of the inner product of the function with itself:

$$norm^2 = \|f(t)\|^2 = \int_{-L}^{L} f(t)\,f*(t)\,dt$$

(11.27)

What is the norm of h(t), the right hand side of Eq. 11-8? Use $L = \pi$.

Exercise 11-6: Very important in nearly all types of engineering, the Fourier Transform of a function f(t) is formed by taking the inner product between the function f(t) and the *Fourier kernel* h(t):

$$h(t) = e^{i\omega t}$$

(11.28)

Letting $L = \pi$, what is the Fourier transform of $f(t) = e^{i\omega t}$?

Exercise 11-7: 0.21 is only one solution of i^i. Let $i = \exp(i(\pi/2 + 2\pi n))$ and find at least two other solutions.

----

[6]George F. Simmons, **Differential Equations with Applications and Historical Notes**, New York: McGraw-Hill, 1972, pp.107-109.

Photo credit:
http://www-gap.dcs.st-and.ac.uk/~history/Mathematicians/Euler.html

Article 12. Higher-order linear differential equations and their IVP's and BVP's

When we solve problems involving higher-order differential equations, we will want to work from the general to the specific: At first we will determine a solution to the differential equation that is as general as possible. Thus we will use just as many free parameters as are needed in order to express every possible manifestation of a solution function to the differential equation. For example, for the differential equation $\ddot{u} + u = 0$, the function $u(t) = C_1\cos(t) + C_2\sin(t)$ is the **general solution**: There is no solution that cannot be expressed by some choice of C_1 and C_2. After the general solution has been determined, we finish solving the problem by using initial conditions or boundary conditions in order to find as many specific values for the C's as possible. Ideally, if things work in our favor, we determine a unique solution. For important types of differential equations, we can determine general solutions by recognizing the set of solutions as a *solution space*--a vector space whose vectors (or points) are solution functions to the differential equation. Then, if the differential equation has the right properties and if we have enough initial or boundary conditions, we can determine the unique point (C_1, C_2) in the solution space that represents the unique solution. To warm up to this idea, we will start by discussing the solution of $\ddot{u} + u = 0$ as a sort of parable for what will be discussed more fully later.

Much of the progress in applied mathematics during the 20th century was due to understanding how to treat functions as vectors. This approach is especially helpful in solving linear differential equations in a general, complete way. We are familiar with the x-y plane. The x-y plane is a two-dimensional space and any point (x_0,y_0) in the plane can be represented as the vector (or the tip of the vector) $x_0\hat{i} + y_0\hat{j}$ where $\hat{i}$ and $\hat{j}$ are basis vectors parallel to the x- and y-axes, respectively. The x-y plane has two **dimensions** and is called **Euclidean two-space**. For example, if $\Re$ is the set of all real numbers, we can understand the point (4,3) as a point in the two-dimensional Euclidean space $\Re \times \Re$ or as the vector $4\hat{i} + 3\hat{j}$. The diagram below illustrates this:

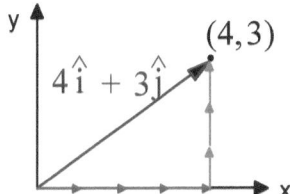
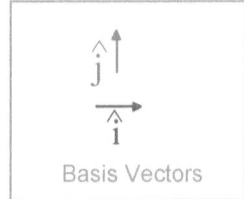

Representation using linearly independent vectors that are orthogonal.

67

In the same way that any point in Euclidean 2-space can be expressed in terms of $\hat{i}$ and $\hat{j}$, any solution to the differential equation $\ddot{u} + u = 0$ can be written as a linear combination of the **basis functions** $u_1 = \cos(t)$ and $u_2 = \sin(t)$. The set of all functions $u = u(t)$ that solve $\ddot{u} + u = 0$ is the **solution space** of $\ddot{u} + u = 0$. Any solution in the solution space of $\ddot{u} + u = 0$ can be represented as a linear combination $u = C_1 u_1 + C_2 u_2 = C_1 \cos(t) + C_2 \sin(t)$. The point (C_1, C_2) in the solution space of $\ddot{u} + u = 0$ plays the same role that the coordinate (x,y) played in the Euclidean two-space. It is a representation of any point in the space; thus it represents any solution of the differential equation. For example, the function $u = 4\cos(t) + 3\sin(t)$ is a solution of $\ddot{u} + u = 0$. In the solution space of $\ddot{u} + u = 0$, the solution function $4\cos(t) + 3\sin(t)$ can be identified with the point $(4,3)$ without ambiguity. Below is a diagram of the solution space:

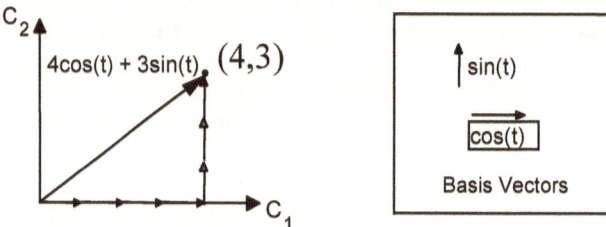

Representation using linearly independent basis functions that are orthogonal

We know that, in Euclidean two-space, we need two linearly independent vectors to serve as basis vectors. The strongest kind of linear independence occurs when the basis vectors are orthogonal; however, we could get by with using $\hat{e}_1 = \hat{i} + \dfrac{1}{2}\hat{j}$ and $\hat{e}_2 = \hat{j}$ since $\hat{e}_1$ and $\hat{e}_2$ in a linear combination can reach any point. For example, the point $(4,3)$ in the x-y plane is the tip of the arrow $4\hat{e}_1 + 1\hat{e}_2$. The vectors $\hat{e}_1$ and $\hat{e}_2$ are linearly independent: You cannot write one of these as a linear combination of the other. But linear independence is good enough to express any vector in the plane in the form $C_1\hat{e}_1 + C_2\hat{e}_2$.

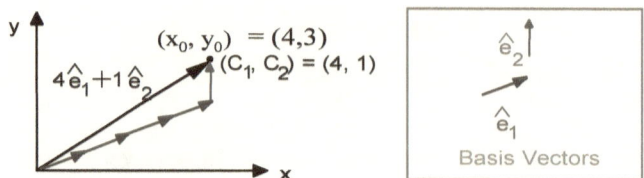

Representation using linearly independent vectors that are not orthogonal

What must be avoided, however, is **linear dependence** which occurs when one or more basis vectors can be expressed as a linear combination of one or more of the other basis vectors and there are only n vectors. For example, the point (4,3) in the x-y plane cannot be reached by:

$$C_1\hat{e}_1 + C_2\hat{e}_2 \text{ if } \hat{e}_1 = \hat{i} \text{ and } \hat{e}_2 = 2\hat{i} \tag{12.1}$$

Since no combination of these can bring you above the x-axis.

Orthogonal vectors have the property that their inner product is equal to zero.

This is true for the basis vectors $\hat{i}$ and $\hat{j}$:

$$\hat{i} \cdot \hat{j} = 0 \tag{12.2}$$

As we saw in Article 11, the inner product between two functions g(t) and h(t) is formed by multiplying the first function times the conjugate of the second and integrating the result. If the inner product is zero then the functions are orthogonal.

For the differential equation ü + u = 0, we can write it in the form ü = - u and ask: "What function(s) can I differentiate twice and obtain the negative of the original?" Sine has that property and so does cosine:

$$\frac{d^2}{dx^2}(\sin t) + \sin t = -\sin t + \sin t = 0$$

$$\frac{d^2}{dx^2}(\cos t) + \cos t = -\cos t + \cos t = 0 \tag{12.3}$$

But sin(t) and cos(t) are also orthogonal functions, since their inner product is zero; therefore, they represent a basis for the solution space of ü + u = 0. Any solution can be represented in the form:

$$u(t) = C_1 \cos(t) + C_2 \sin(t) \tag{12.4}$$

It is no coincidence that, for this 2nd-order differential equation, we need *two* linearly independent basis functions in order to express the general solution. The basis functions cos(t) and sin(t) are not only linearly independent; they are also orthogonal. We could represent the general solution in terms of two basis functions that are linearly independent but *not* orthogonal, say u(t) = $C_1e_1(t)$ + $C_2e_2(t)$ where $e_1(t)$ = cos(t) + ¼sin(t) and $e_2(t)$ = ¾sin(t). Even so, we will usually find it convenient, if possible, to find orthogonal basis functions.

We are ready now to talk about higher-order differential equations in greater generality. By *higher-order* we mean n[th]-order (also said "n-order") with *n* greater than one.

Definition 12-1: An nth-order linear ordinary differential equation, on the interval I, is an equation of the form:

$$a_n(x)\frac{d^n y}{dx^n} + a_{n-1}(x)\frac{d^{n-1} y}{dx^{n-1}} + ... + a_1(x)\frac{dy}{dx} + a_0(x)y = g(x)$$

$$(12.5)$$

Such equations are *linear* in the sense that the highest order of derivative of the dependent variable may be written as a linear function of any lower-order derivatives of the dependent variable:

$$\frac{d^n y}{dx^n} = -\frac{a_{n-1}(x)}{a_n(x)}\frac{d^{n-1} y}{dx^{n-1}} - \frac{a_{n-2}(x)}{a_n(x)}\frac{d^{n-2} y}{dx^{n-2}} - ... - \frac{a_0(x)}{a_n(x)}y$$

$$= \frac{g(x)}{a_n(x)}$$

$$(12.6)$$

To comprehend this, it helps to think of $-\dfrac{a_{n-1}(x)}{a_n(x)}$ as $m_1(x)$ and

$-\dfrac{a_{n-2}(x)}{a_n(x)}$ as $m_2(x)$ in harkening back to the old "(dependent variable) = m(independent variable) + b" concept of "linear" from our age of innocence. If you combine Eq. 12.5 with the initial conditions

$$y(x_0) = y_0, \quad y'(x_0) = y_1, \quad ..., \quad y^{(n-1)}(x_0) = y_{n-1} \quad (12.7)$$

Then you have a linear n[th]-order *initial-value problem*. For such problems, the solution (on some interval I of the independent variable) is **unique** if all of the "*a* functions" as well as the function g are continuous and if:

$$a_n(x) \neq 0 \text{ for all x} \in \text{I} \tag{12.8}$$

Example 12-1: Solve the following IVP:

$$y'' + y = 0 \text{ with initial conditions } y(0) = 0 \text{ and } y'(0) = 1 \tag{12.9}$$

The solution is going to be unique since $a_2(x)$, the leading "*a* function," does not equal zero and the other *a*'s and *g* are continuous. But first let's write down the general solution and then we'll determine the specific, unique solution to the given problem. As we have seen, sine and cosine are each solutions to the differential equation:

$$\frac{d^2}{dx^2}(\sin x) + \sin x = -\sin x + \sin x = 0$$

$$\frac{d^2}{dx^2}(\cos x) + \cos x = -\cos x + \cos x = 0 \tag{12.10}$$

The linear combination $C_1cos(x) + C_2sin(x)$ also provides a solution:

$$\frac{d^2}{dx^2}(C_1 \cos x + C_2 \sin x) + C_1 \cos x + C_2 \sin x = \tag{12.11}$$

$$-C_1 \cos x + -C_2 \sin x + C_1 \cos x + C_2 \sin x = 0 \tag{12.12}$$

Indeed $C_1cos(x) + C_2sin(x)$ is the **general solution** to Eq. 12-9: it is a vector that reaches any point in the solution space of $y'' + y = 0$ and (C_1, C_2) can be used to represent any point in the solution space. So now we have only to apply the initial conditions and determine exactly which unique point (C_1, C_2) in the solution space is the ultimate solution to the problem. Taking

$$y = C_1 \cos(x) + C_2 \sin(x) \tag{12.13}$$

and plugging in the initial conditions, we have:

$$y(0) = C_1 \cos(0) + C_2 \sin(0) = C_1 = 0$$

$$y'(0) = -C_1 \sin(0) + C_2 \cos(0) = C_2 = 1 \tag{12.14}$$

Thus we have been able to determine C_1 and C_2; therefore, the final unique solution to the present problem is:

$$y = \sin(x) \tag{12.15}$$

It is the point $(C_1, C_2) = (0,1)$ in the solution space with basis vectors $\hat{e}_1 = \cos(x)$, $\hat{e}_2 = \sin(x)$.

◊

The harmonic oscillator example we have been using has the general solution with the auspicious vector space properties for more reasons than one. In addition to being a linear differential equation, it is also a homogeneous equation.

Definition 12-2: A **homogeneous linear differential equation** is one for which $g(x) = 0$ in Eq. 12-4:

$$a_n(x)\frac{d^n y}{dx^n} + a_{n-1}(x)\frac{d^{n-1} y}{dx^{n-1}} + \dots + a_1(x)\frac{dy}{dx} + a_0(x)y = 0 \tag{12.16}$$

This homogeneity of the equation has nothing to do with the "homogeneous coefficients" $M(x, y)$ and $N(x, y)$ that we discussed previously. Here homogeneous means that if y is a solution, then Cy is a solution. (You can look at the equation and see that C divides out.) Also, if y_1 and y_2 are solutions to an equation, then $C_1 y_1 + C_2 y_2$ is also a solution. That's why $C_1 y_1 + C_2 y_2$ is the "general solution" of the homogeneous ODE.

A solution space of a differential equation is the set of all possible solutions of that equation. A linear homogeneous ODE creates a **solution space** that has the same dimension as the order of the equation and the same number of basis functions as that order. The nth-order homogeneous linear ODE has its own n-dimensional solution space and n functions serving as basis functions. As we have alluded to already, there is a superposition principle at work when we write a solution as a linear combination of these n functions. If y_1, y_2, . . . y_k are solutions to :

$$a_k(x)\frac{d^k y}{dx^k} + a_{k-1}(x)\frac{d^{k-1} y}{dx^{k-1}} + \dots + a_1(x)\frac{dy}{dx} + a_0(x)y = 0 \tag{12.17}$$

Then

72

$$y = C_1 y_1 + C_2 y_2 + ... + C_k y_k \qquad (12.18)$$

is also a solution.

If y_1, y_2, ... y_n are independent then they can be used as a **basis for the solution space** of the homogeneous equation:

$$a_n(x)\frac{d^n y}{dx^n} + a_{n-1}(x)\frac{d^{n-1} y}{dx^{n-1}} + ... + a_1(x)\frac{dy}{dx} + a_0(x)y = 0$$

$$(12.19)$$

Then the y_1, y_2, ... y_n are independent basis vectors of the space. The basis { y_1, y_2, ... y_n } forms a coordinate system in which any solution $y = f(x)$ can be expressed:

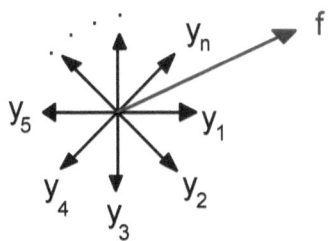

The basis vectors {y_1, y_2,...,y_n} can be tested for linear independence by calculating the Wronskian W, which is the following determinant:

$$W = \begin{vmatrix} y_1 & y_2 & \cdots & y_n \\ y_1' & y_2' & \cdots & y_n' \\ \vdots & \vdots & & \vdots \\ y_1'' & y_2'' & \cdots & y_n'' \end{vmatrix}$$

$$(12.20)$$

They are linearly independent if, for some point x_0 in I, W does not equal zero. Thus the condition for linear independence is:

For $\{y_1, y_2, ..., y_n\}$ defined on I, if $W \neq 0$ for some $x_0 \in I$
then $\{y_1, y_2, ..., y_n\}$ are linearly independent. $\qquad$ 12.21)

Example 12-2: Show that the functions {sin(x), cos(x)} are linearly independent by calculating the Wronskian. We have:

$$W = \begin{vmatrix} \sin(x) & \cos(x) \\ \cos(x) & -\sin(x) \end{vmatrix} = -\sin^2(x) - \cos^2(x) = -1 \neq 0$$

(12.22)

The Wronskian is not equal to zero; therefore, {sin(x), cos(x)} are linearly independent.

◊

The quintessential example of a boundary value problem is an elastic string fixed at both ends. There are boundary conditions on the motion of the string (the fact that the walls are boundaries and so the string can't move up and down at the walls).

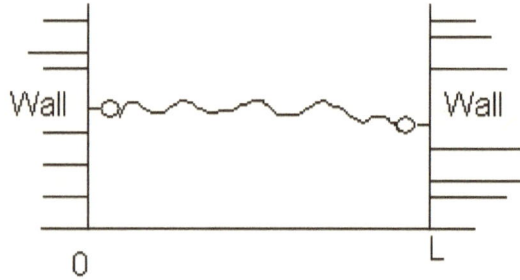

A **boundary value problem (BVP)** is a differential equation with boundary conditions. For the string, the walls provide boundaries which "contain" the motion, since a wave on the string cannot continue into the wall (except by a very small amount). The physical system above can be modeled using the following BVP:

$$y'' + \omega_m^2 \, y = 0$$

With boundary conditions y(0) = 0, y(L) = 0.   (12.23)

The subscript m appearing on the parameter omega is to indicate that there will be harmonics. (Just like a guitar string). This is really a whole bunch of different equations…one for every different value of m. The general solution is:

$$y = C_1 \cos(\omega_m x) + C_2 \sin(\omega_m x)$$

(12.24)

When we plug in the first boundary condition we obtain:

$$y(0) = C_1 = 0$$

$$(12.25)$$

which kills off the first term right of the equals sign in equation Eq. 12-24. Plugging in the second boundary condition, we see that:

$$y(L) = C_2 \sin(\omega_m L) = 0,$$

which requires that $\omega_m L = q\pi, \quad q = ...,-2,-1,0,1,2,...$ (12.26)

So the solutions are:

$$y(x) = C \sin\left(\frac{q\pi}{L}x\right)$$

$$(12.27)$$

The solutions look like this for some different values of q:

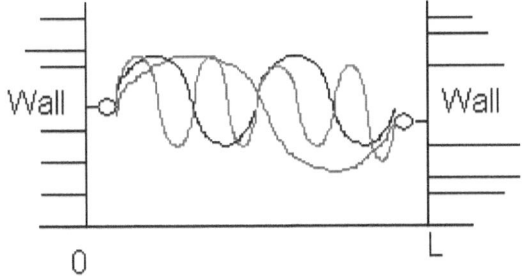

So far we've talked about higher-order linear ODE's that are homogeneous (meaning $g(x) = 0$). What we've learned so far about homogeneous equations is a great leg-up on solving the non-homogeneous equation:

$$a_n(x)\frac{d^n y}{dx^n} + a_{n-1}(x)\frac{d^{n-1} y}{dx^{n-1}} + ... + a_1(x)\frac{dy}{dx} + a_0(x)y = g(x)$$

$$(12.28)$$

In order to solve the non-homogeneous equation, we first find the general solution $C_1y_1+C_2y_2+...+C_ny_n$ to the homogeneous equation formed by setting $g(x) = 0$. (This will be called the homogeneous solution y_h.) Then we find one particular solution y_p to the *full* non-homogeneous solution and add it to the homogeneous solution y_h to form the general solution y_g: The general solution to the non-homogeneous equation then is:

$$y_g = y_h + y_p \tag{12.29}$$

Or:

$$y_g = C_1 y_1 + C_2 y_2 + ... + C_n y_n + y_p \tag{12.30}$$

Thus an algorithm for solving the non-homogeneous linear ODE, Eq. 12-28, is:

1.) Set $g(x) = 0$, solve the homogeneous equation and establish the homogeneous part of the solution $y_h = C_1 y_1 + ... + C_n y_n$.
2.) Find a particular solution y_p for the non-homogeneous case.
3.) Then the general solution of the overall non-homogeneous problem is $y_g = y_h + y_p$.

Let's apply this algorithm to an example. At the same time, we will try out a new technique for solving homogeneous linear ODE's with constant coefficients. For such equations, we look for solutions in the form $y = e^{ax}$. We will explore this method more fully in the next article.

Example 12-3: Solve:

$$y'' - 4y = 6x - 4x^3 \tag{12.31}$$

Solution: The first step is to set $g(x) = 0$ and solve the corresponding homogeneous equation:

$$y'' - 4y = 0 \tag{12.32}$$

I need two independent solutions since $y'' - 4y = 0$ is 2nd order. (These two will form the basis for the solution space of the homogeneous equation.) Look for solutions in the form $y = e^{ax}$. The second derivative is $y'' = a^2 e^{ax}$. Plugging this in we get:

$$y'' - 4y = a^2 e^{ax} - 4e^{ax} = 0 \tag{12.33}$$

Dividing out e^{ax}, we obtain the **auxiliary** equation:

$$a^2 - 4 = 0 \tag{12.34}$$

Which gives us two values for a:

$$a = \pm 2 \qquad (12.35)$$

Since, by looking for solutions in the form $y = e^{ax}$, we came up with a $= \pm 2$, we have two solutions to the homogeneous equation: $y_1 = e^{2x}$ and $y_2 = e^{-2x}$. We can show that these are independent by demonstrating a non-zero Wronskian:

$$W = \begin{vmatrix} e^{2x} & e^{-2x} \\ 2e^{2x} & -2e^{-2x} \end{vmatrix} = -2 - 2 = -4 \neq 0 \qquad (12.36)$$

So the homogeneous solution y_h is:

$$y_h = C_1 e^{2x} + C_2 e^{-2x} \qquad (12.37)$$

Now we have only to find one single particular solution y_p, then we can add this to the homogeneous solution and finish solving the non-homogeneous differential equation. For the time being, we will find y_p by guessing. Let's try $y_p = x^3$ since it's one of the terms in $g(x)$. Plugging in we see that:

$$y_p'' - 4y_p = 6x - 4x^3 \qquad (12.38)$$

Thus the non-homogeneous equation is solved by this particular solution. (We will soon learn how to determine particular solutions without guessing.) The answer to the original problem then is the general solution $y_g = y_h + y_p$ which is the sum of the homogeneous and the particular solution:

$$y = y_h + y_p = C_1 e^{2x} + C_2 e^{-2x} + x^3 \qquad (12.39)$$

Here no boundary or initial conditions are presented in the problem statement; therefore, we are done.

◊

The proof that $y_h + y_p$ is a solution to Eq. 12-28 is one of the easiest proofs in differential equations. If y_h is the homogeneous solution, then:

$$a_n(x)\frac{d^n y_h}{dx^n} + \ldots + a_1(x)\frac{dy_h}{dx} + a_0(x)y_h = 0 \qquad (12.40)$$

If y_p is a particular solution, then

$$a_n(x)\frac{d^n y_p}{dx^n} + \ldots + a_1(x)\frac{dy_p}{dx} + a_0(x)y_p = g(x)$$

$$(12.41)$$

If we add Eq. 12-39 and Eq. 12-40 together, we obtain:

$$a_n(x)\frac{d^n(y_h + y_p)}{dx^n} + \ldots + a_1(x)\frac{d(y_h + y_p)}{dx} + a_0(x)(y_h + y_p) =$$

$$g(x)$$

$$(12.42)$$

So we have proved that $y_h + y_p$ must also be a solution.

___

EXERCISES

Exercise 12-1:

Three solutions of $\ddot{y} + \omega^2 y = 0$ are:

$$\left\{ \sin(\omega t),\ \cos(\omega t),\ \sqrt{2}\cos\left(\omega t - \frac{\pi}{4}\right) \right\}.$$

Show that these are not linearly independent by writing the third function as a linear combination of the first two.

Exercise 12-2: Calculate the Wronskian for the functions of the previous example:

$$W = \begin{vmatrix} y_1 & y_2 & y_3 \\ y_1' & y_2' & y_3' \\ y_1'' & y_2'' & y_3'' \end{vmatrix}$$

$$(12.43)$$

What is your conclusion?

Exercise 12-3: A very simplified radio consists of an inductor L and a capacitor C. It is picking up a weak radio signal $\sin(\omega_0 t)$ at frequency ω_0.

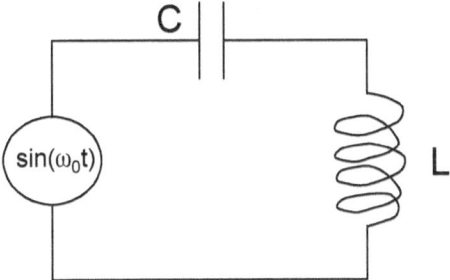

The current j(t) is the rate at which charges Q are flowing per unit time. The voltage drop across the inductor is L(dj/dt) and the voltage drop across the capacitor is Q/C. Kirchoff's Law says that these voltage drops must add up to the applied voltage:

$$L\frac{dj}{dt} + \frac{1}{C}Q = \sin(\omega_0 t)$$

$$(12.44)$$

The current j is the time derivative of charge: $j(t) = dQ/dt$. Differentiating Eq. 12-44 we obtain a second order differential equation:

$$L\frac{d^2 j}{dt^2} + \frac{1}{C}j = \omega_0 \cos(\omega_0 t)$$

$$(12.45)$$

Which can be placed in the form of Eq. 12-5 by dividing through by L:

$$\frac{d^2 j}{dt^2} + \frac{1}{LC}j = \frac{\omega_0}{L}\cos(\omega_0 t)$$

$$(12.46)$$

Show that the homogeneous solution is:

$$j_h = C_1 \cos(\frac{1}{\sqrt{LC}}t) + C_2 \sin(\frac{1}{\sqrt{LC}}t)$$

$$(12.47)$$

Exercise 12-4: Show that a particular solution to the differential equation of the previous problem is:

$$j_p = \frac{\omega_0 C}{1 - \omega_0^2 LC}\cos(\omega_0 t)$$

(12.48)

Exercise 12-5: Write down the general solution to Eq. 12-45.

Exercise 12-6: To make a more realistic radio, we can add a resistor R.

$$L\frac{d^2 j}{dt^2} + R\frac{dj}{dt} + \frac{1}{C}j = \omega_0 \cos(\omega_0 t)$$

(12.49)

A particular solution to this is:

$$j_p = \frac{1}{\sqrt{R^2 + \left[\omega_0 L - \dfrac{1}{\omega_0 C}\right]^2}}\cos(\omega_0 t)$$

(12.50)

Note that this reduces to Eq. 12-46 when R = 0. Suppose the radio is now made tunable by providing for the capacitor to have adjustable capacitance C. What value of capacitance C provides the largest amplitude for y_p? (This **electrical resonance** is the method by which radios tune to channels.)

Exercise 12-7: A guitarist plucks a guitar string which has a length of L between places where it is tightly fixed at x = 0 and x = L.

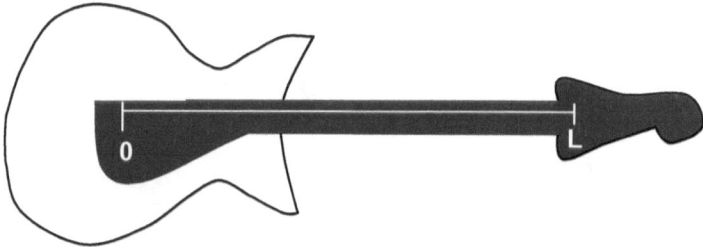

Assuming the same boundary value problem as Eq. 12-22, what is the lowest non-zero frequency that can be heard?

Exercise 12-8: If the guitarist places his finger at x = L/2, and plucks to the right of that, what is the lowest non-zero frequency that can be heard.

Article 13. Homogeneous linear ODE's with constant coefficients

A homogeneous linear equation with constant coefficients is an equation in the form:

$$a_n y^{(n)} + a_{n-1} y^{(n-1)} + ... + a_1 y' + a_0 y = 0 \qquad (13.1)$$

Where all the a's are constants. All solutions are either exponentials or they are made up of exponentials, possibly complex exponentials. So the approach here is to look for solutions of the form $y = e^{mx}$. This results in a polynomial equation in m which we then try to solve for the roots.

Example 13-1:   Solve:

$$y'' - 5y' + 6y = 0 \qquad (13.2)$$

Letting $y = e^{mx}$ we obtain $y' = me^{mx}$ and $y'' = m^2 e^{mx}$ (13.3)

Plugging in, we have:

$$y'' - 5y' + 6y = m^2 e^{mx} - 5me^{mx} + 6e^{mx} = 0 \qquad (13.4)$$

We can divide out e^{mx} and what we're left with is called the "characteristic equation" or the "auxiliary equation."

$$m^2 - 5m + 6 = 0 \qquad (13.5)$$

This can be factored as $(m - 2)(m - 3) = 0$ or $m_1 = 2$ and $m_2 = 3$. So, by assuming $y = e^{mx}$ we have arrived at two solutions e^{2x} and e^{3x}. The general solution then is:

$$y = C_1 e^{2x} + C_2 e^{3x} \qquad (13.6)$$

◊

When using this $y = e^{mx}$ trick for solving second-order equations, that is, equations of the form

$$ay'' + by' + cy = 0 \qquad (13.7)$$

There are three things that can happen:

1.) if $b^2 - 4ac > 0$ then m_1 and m_2 are real and distinct.
2.) if $b^2 - 4ac = 0$ then $m_1 = m_2$ (we have a double root).
3.) if b2 – 4ac < 0 then m_1 and m_2 are complex conjugate (meaning that if, say, $m_1 = 2i$, then $m_2 = -2i$).

Case one is the most simple; we obtain two distinct m's this gives us two independent functions and we have what we need to write the general solution. Case two gives me only one solution; therefore, we have to use some other technique (for instance **reduction of order**) to get the other solution. (We will discuss reduction of order more fully later on. One means by which to implement reduction of order involves looking for a second solution $y_2(x)$ as the product of the first solution $y_1(x)$ and some function u(x). That is, plug in $y_2(x) = u(x)y_1(x)$.) Case three might be reducible to real using Euler's formula:

$$e^{ix} = \cos(x) + i\sin(x) \qquad (13.8)$$

Example 13-1 was an example of Case 1: The auxiliary equation has two distinct, real roots. Now let's look at a Case 2 example, an example in which the auxiliary equation has a double root:

Example 13-2: Case 2: A double root. Solve:

$$y'' - 6y' + 9y = 0 \qquad (13.9)$$

Our trial solution is $y = e^{mx}$. The derivatives of this are $y' = me^{mx}$ and $y'' = m^2 e^{mx}$. Plugging these into Eq. 13-9, we obtain:

$$m^2 e^{mx} - 6me^{mx} + 9e^{mx} = 0 \qquad (13.10)$$

Which gives the characteristic equation:

$$m^2 - 6m + 9 = 0 \quad \text{which factors as} \quad (m - 3)^2 = 0 \quad (13.11)$$

This gives the double root m = 3. So we only obtain one basis function: $y_1 = e^{3x}$. But we need two basis functions y_1 *and* y_2 to express the general solution of a second-order linear ODE. Now use reduction of order to obtain the second solution y_2. We look for a second solution in the form $y_2 = u(x)y_1$. Here we seek a solution in the form:

$$y_2 = u(x)e^{3x} \qquad (13.12)$$

Then:

$$y_2' = u'e^{3x} + 3u^{3x}$$

$$y_2'' = u''e^{3x} + 3u'e^{3x} + 3u'e^{3x} + 9ue^{3x}$$

(13.13)

So, plugging into Eq. 13-9, we get:

$$u''e^{3x} + 6u'e^{3x} + 9ue^{3x} - 6u'e^{3x} - 18ue^{3x} + 9ue^{3x} = 0$$

(13.14)

Lucky for us, all of this collapses down to a succinct little equation:

$$u''e^{3x} = 0 \quad \text{or:} \quad u'' = 0$$

(13.15)

Integrating, we are able to find a second solution:

$$u = C_1 x + C_2$$

(13.16)

Thus:

$$y = ue^{3x} = C_1 xe^{3x} + C_2 e^{3x}$$

(13.17)

We can show that the basis functions $\{xe^{3x}, e^{3x}\}$ are linearly independent by verifying that the Wronskian is not zero:

$$W = \begin{vmatrix} xe^{3x} & e^{3x} \\ \left(e^{3x} + 3xe^{3x}\right) & 3e^{3x} \end{vmatrix} = 3xe^{6x} - e^{6x} - 3xe^{6x} = -e^{6x}$$

(13.18)

There is no real x that will make $-e^{6x}$ equal to zero; therefore $\{xe^{3x}, e^{3x}\}$ is a basis for the solution space of Eq. 13-9 and Eq. 13-17 is the general solution.

◊

Example 13-3: Case 3: Complex conjugate roots. Solve the harmonic oscillator:

$$y'' + y = 0$$

(13.19)

Again, our trial solution is $y = e^{mx}$. The second derivative of this is $y'' = m^2 e^{mx}$. Plugging these in, we obtain:

$$m^2 e^{mx} + 1e^{mx} = 0 \quad \text{which gives} \quad m^2 + 1 = 0 \quad (13.20)$$

The solution to the auxiliary equation is a couple of complex conjugate roots:

$$m = \pm i \quad (13.21)$$

Thus, the solution this gives us is:

$$y = C_1 e^{ix} + C_2 e^{-ix} \quad (13.22)$$

Using Euler's formula we have:

$$y = C_1(\cos(x) + i\sin(x)) + C_2(\cos(x) - i\sin(x))$$
$$y = (C_1 + C_2)\cos x + i(C_1 - C_2)\sin x \quad (13.23)$$

We are only interested in solutions to this which are real. Taking the real component of this, we have $y_1 = \cos(x)$. We can use reduction of order to get another linearly independent solution y_2 in order to be able to write the general solution. Let $y_2 = u(x)\cos(x)$. The second derivative of this is $y_2'' = u''\cos(x) - 2u'\sin(x) - u\cos(x)$. Plugging into Eq. 13-19, we have:

$$u''\cos(x) - 2u'\sin(x) - u(x)\cos(x) + u(x)\cos(x) = 0 \quad (13.24)$$

Notice that the terms involving u(x) cancel each other. Now define $w(x) = u'$ in order to reduce order to first-order:

$$w'\cos(x) - 2\sin(x)w = 0 \quad (13.25)$$

We can use separable variables to solve for w:

$$\frac{dw}{w} - 2\tan(x)\,dx = 0 \quad (13.26)$$

Integrating, we have:

$$\ln(w) + 2\ln(\cos(x)) = \ln C \quad (13.27)$$

Which may be written as:

$$w = C_1 \sec^2 x = u'$$ (13.28)

Integrating, we determine u:

$$u(x) = C_1 \tan x + C_2$$ (13.29)

We said that the second solution is the product of u(x) and the first solution cos(x):

$$y_2(x) = u(x) y_1(x)$$
$$= (C_1 \tan x + C_2) \cos x = C_1 \sin x + C_2 \cos x$$ (13.30)

As we've shown before, {sinx, cosx} is a linearly independent set of basis functions for the solution space; therefore, we have what we need to present the general solution:

$$y(x) = C_1 \sin x + C_2 \cos x$$ (13.31)

◊

As the previous example showed, the case of complex conjugate roots can be a chore. To simplify matters, we have the following theorem:

Theorem:

If the auxiliary equation has distinct complex conjugate roots m = $\alpha \pm \beta i$, then the general solution to y″ + by′ + cy = 0 is:

$$y = e^{\alpha x}(C_1 \cos(\beta x) + C_2 \sin(\beta x))$$ (13.32)

Example 13-4: Solve:

$$y'' - 14y' + 74y = 0$$ (13.33)

Plugging in y = e^{mx} and the derivatives y′ = me^{mx} and y″ = $m^2 e^{mx}$, we have, after dividing out the common term e^{mx}:

$$m^2 - 14m + 74 = 0$$ (13.34)

Shortcut to Ordinary Differential Equations

The quadratic formula gives us:

$$m = 7 \pm i5 = \alpha \pm i\beta \tag{13.35}$$

Thus the general solution is:

$$y = e^{7x}(C_1 \cos(5x) + C_2 \sin(5x)) \tag{13.36}$$

◊

Be careful to keep in mind that the above theorem only applies to constant coefficient, second-order linear ODE's.

We can summarize the situation for a 2nd-order linear ODE with constant coefficients in a convenient table as follows.

Table 13-1: *Homogeneous solutions for 2nd-order linear ODE with constant coefficients.*

Case 1	Distinct real roots	$y_h = C_1 e^{m_1 x} + C_2 e^{m_2 x}$
Case 2	Repeated real root	$y_h = C_1 e^{mx} + C_2 x e^{mx}$
Case 3	Complex roots	$y_h = e^{\alpha x}[C_1 \cos(\beta x) + C_2 \sin(\beta x)]$

EXERCISES

Exercise 13-1: Find the general solution of:

$$\frac{d^2 y}{dx^2} - 3\frac{dy}{dx} + 2y = 0 \tag{13.37}$$

Exercise 13-2: Find the general solution of:

$$\ddot{y} - 2\dot{y} + 5y = 0 \tag{13.38}$$

Exercise 13-3: Find the general solution of:

$$y'' - 8y' + 16y = 0 \tag{13.39}$$

Exercise 13-4: Solve the IVP:

$$y'' - (2 + \pi)y' + 2\pi = 0 \qquad (13.40)$$

With initial conditions y(0) = 1 and y'(0).

Article 14. Reduction of order

In the previous article, we used reduction of order to find a second solution to a differential equation whose auxiliary equation had a double root. This was accomplished by making the substitution $y_2(x) = u(x)y_1(x)$, a method that Euler originated. It turns out that there are several techniques that accomplish reduction of order involving several different types of substitutions. The problem at hand determines which approach to use.

The simplest reduction of order technique involves making the substitution $u(x) = y'$. An obvious occasion to use this is when y does not appear explicitly in a differential equation:

<u>Reduction of order algorithm 1:</u> Consider the 2nd-Order Linear DE:

$$a_2(x)y'' + a_1(x)y' + a_0(x)y = b(x) \qquad (14.1)$$

If $a_0(x) = 0$ then we can reduce the order by simply setting:

$$u = y' \qquad (14.2)$$

•

This substitution reduces the equation to

$$a_2(x)u' + a_1(x)u = b(x) \qquad (14.3)$$

which is linear, first-order and can be solved using techniques we know already.

Example 14-1: Solve:

$$xy'' + y' = 0 \qquad (14.4)$$

Solution:

$$u = y' \quad \text{and} \quad u' = y'' \qquad (14.5)$$

Thus the equation reduces to:

$$xu' + u = 0 \qquad (14.6)$$

Or:

$$x \frac{du}{dx} = -u \tag{14.7}$$

$$\frac{du}{u} = -\frac{dx}{x} \tag{14.8}$$

This integrates to become:

$$\ln u = -\ln x + \ln C \tag{14.9}$$

$$u = \frac{C}{x} \tag{14.10}$$

Substituting u = y′ back in, we have:

$$y' = \frac{C}{x} \tag{14.11}$$

Thus the general solution is:

$$y = C_1 \ln x + C_2 \tag{14.12}$$

$$\Diamond$$

A method also called reduction of order, which we have some experience with from Article 13, involves making the substitution:

$$y_2(x) = u(x)y_1(x) \tag{14.13}$$

To reduce the order. This is a handy tool when we already have one solution $y_1(x)$ and we are seeking a second solution $y_2(x)$

Reduction of order algorithm 2:

If we have:

$$a_2(x)y'' + a_1(x)y' + a_0(x)y = 0 \tag{14.14}$$

And if we have a non-trivial solution $y_1(x)$, then we can find a second, linearly independent solution $y_2(x)$ by substituting:

$$y_2(x) = u(x)y_1(x) \qquad (14.15)$$

This substitution causes the 2nd-order DE to reduce to a first order. \

Example 14-2: Find the general solution to the following differential equation, given one solution $y_1(x)$:

$$y'' - 4y' + 4y = 0 \qquad \text{has solution} \qquad y_1 = e^{2x} \qquad (14.16)$$

Solution:

We form $y_2 = uy_1$ and differentiate:

$$y_2' = u'y_1 + uy_1' = u'e^{2x} + u2e^{2x} \qquad (14.17)$$

Differentiating again, we get:

$$y_2'' = e^{2x}u'' + 4e^{2x}u' + 4e^{2x}u \qquad (14.18)$$

Plugging y_2, y_2', and y_2'' into the differential equation we obtain:

$$e^{2x}u'' + 4e^{2x}u' + 4e^{2x}u - 4u'e^{2x} - 8ue^{2x} + 4ue^{2x} = 0 \qquad (14.19)$$

Which reduces to:

$$u'' = 0 \qquad (14.20)$$

Integrating twice we obtain:

$$u = C_1x + C_2 \qquad (14.21)$$

Thus the second solution we are looking for is:

$$y_2 = uy_1 = C_1e^{2x}x + C_2e^{2x} \qquad (14.22)$$

The functions $\{xe^{2x}, e^{2x}\}$ are linearly independent:

$$W = \begin{vmatrix} xe^{2x} & e^{2x} \\ e^{2x}(1+2x) & 2e^{2x} \end{vmatrix} = 2xe^{4x} - (1+2x)e^{4x} = e^{4x}$$

(14.23)

This is owing to the fact that the Wronskian is not zero. That being the case, the general solution is:

$$y = C_1 xe^{2x} + C_2 e^{2x}$$

(14.24)

◊

      Let us understand the linear algebra in the second algorithm. We have a 2nd order linear DE here so we can expect the solution space to be spanned by two solutions y_1 and y_2 that are independent. We are given y_1. What does it mean to look for the other solution to be $y_2 = u(x)y_1$? This is a very important point. The answer is that, by setting $y_2 = u(x)y_1$, we are coercing the answer to not be $y_2 = Cy_1$, which is a simple multiple of y_1. Again, by forcing the two solutions not to be multiples of each other we assure they are not simple multiples of each other.

      There is one other method known as reduction of order which is important to mention. This occurs when the independent variable x does not explicitly appear. In such cases, a subtle chain rule trick may be used to reduce order.

Reduction of order algorithm 3:

If we have the constant coefficient 2nd-order equation:

$$a_2 y'' + a_1 y' + a_0 y = b$$

(14.25)

Then, defining:

$$u = \frac{dy}{dx}$$

(14.26)

We note that:

$$y'' = u' = \frac{du}{dx} = \frac{du}{dy}\frac{dy}{dx} = u\frac{du}{dy} \tag{14.27}$$

Replacing y' and y'' using these last two results, Eq. 14-25 becomes first-order:

$$a_2 u \frac{du}{dy} + a_1 u + a_0 y = b \tag{14.28}$$

•

Example 14-3: Use reduction of order to solve $y'' + y = 0$.

Solution: Making the substitution for y'' provided by Eq. 14-27, we have:

$$u\frac{du}{dy} + y = 0 \tag{14.29}$$

Multiplying through by dy and integrating, we have:

$$\frac{u^2}{2} + \frac{y^2}{2} = \frac{C_1}{2} \tag{14.30}$$

Thus we can solve for u and re-substitute:

$$u = \frac{dy}{dx} = \sqrt{C_1 - y^2} \tag{14.31}$$

Separating variables, and factoring out C_1 from the radical, we obtain:

$$\frac{dy}{\sqrt{C_1}\sqrt{1 - \left(\dfrac{y}{\sqrt{C_1}}\right)^2}} = dx \tag{14.32}$$

The left side integrates to become the arcsine:

$$\arcsin \frac{y}{\sqrt{C_1}} = x + C_2 \tag{14.33}$$

Taking the sine of each side, and multiplying by the constant, we have, after reducing the sine of a sum:

$$y = \left(\sqrt{C_1}\cos(C_2)\right)\sin(x) + \left(\sqrt{C_1}\sin(C_2)\right)\cos(x) \qquad (14.34)$$

The coefficients $\sqrt{C_1}\cos(C_2)$ and $\sqrt{C_1}\sin(C_2)$ may be used to represent any two real numbers; therefore, we are justified in presenting the solution as:

$$y = C_1\sin(x) + C_2\cos(x) \qquad (14.35)$$

◊

EXERCISES

Exercise 14-1: For the non-linear second-order equation $y'' + y'y = 0$, find two particular solutions y_1 and y_2 by using the substitutions of the third algorithm:

$$y' = u \quad \text{and} \quad y'' = u\frac{du}{dy} \qquad (14.36)$$

Are linear combinations $C_1y_1 + C_2y_2$ also solutions to $y'' + y'y = 0$? If not, why not?

Exercise 14-2: Find the general solution to:

$$y'' - y' + \frac{1}{4}y = 0 \qquad (14.37)$$

Exercise 14-3: Find a particular solution to:

$$xy'' + 2y' = x^2 \qquad (14.38)$$

Article 15. Undetermined coefficients

The non-homogeneous linear ODE with constant coefficients is:

$$a_n \frac{d^n y}{dx^n} + a_{n-1} \frac{d^{n-1} y}{dx^{n-1}} + \ldots + a_1 \frac{dy}{dx} + a_0 y = g(x) \qquad (15.1)$$

For this equation, our ultimate objective is to find the general solution $y_g = y_h + y_p$. We already know how to find a homogeneous solution y_h to the corresponding homogeneous equation $a_n y^{(n)} + a_{n-1} y^{(n-1)} + \ldots + a_1 y' + a_0 y = 0$. The homogeneous solution is a linear combination of linearly independent basis vectors $\{y_1, y_2, \ldots, y_n\}$. The method of undetermined coefficients is helpful for finding a particular solution y_p to the *non-homogeneous* linear ODE with constant coefficients.

The method of undetermined coefficients involves developing a set of candidate component functions $\Im = \{f_1, f_2, \ldots, f_m\}$ that y_p can be constructed from by superposition. Once this set of candidate functions has been gathered together, y_p is written as a superposition of these functions. The A's are the "undetermined coefficients." Thus we write:

$$y_p = A_1 f_1 + A_2 f_2 + \ldots + A_m f_m$$

$$(15.2)$$

By plugging this y_p into the non-homogeneous equation, Eq. 15-1, the A's are determined.

In order to apply the method of undetermined coefficients, $g(x)$ must consist of polynomials, exponentials, sines, and cosines in linear combinations. The reason for this is that these functions are "closed under differentiation." If you differentiate sines and cosines, you get sines and cosines (up to a multiplicative constant). The same is true for exponentials and polynomials.

In building the set of candidate component functions $\Im = \{f_1, f_2, \ldots, f_m\}$, one has to be mindful of the contents and characteristics of both the left- and the right-hand sides of Eq. 15-1; therefore, the homogeneous functions $\{y_1, y_2, \ldots, y_n\}$ also come into play in this algorithm. (For our purposes, we will provide for $\{y_1, y_2, \ldots, y_n\}$ to have constant multipliers divided off and common summands subtracted off. Thus, for example, instead of $\{y_1, y_2\} = \{3\sin x, 2x + \sin x\}$ we will use $\{y_1, y_2\} = \{\sin x, x\}$.) The non-homogeneous term $g(x)$ can be expressed as the sum of its summands: $g = g_1 + g_2 + \ldots$. We will gather these summands together in a set $\Im_1$ and transform and expand this set until we have the set of functions $\mathfrak{I}$ we need to represent all of the possibilities for expressing y_p. The undetermined coefficients algorithm (superposition approach) goes as follows. There are five steps:

Undetermined coefficients algorithm:

1.) The first step toward developing $\mathfrak{I}$ is to include all of the summands of g in a set $\mathfrak{I}_1$:

$$h_1 \leftarrow g_1, \; h_2 \leftarrow g_2, \; ...$$
$$\mathfrak{I}_1 = \{h_1, h_2, ...\} \tag{15.3}$$

2.) Next, we modify the set $\mathfrak{I}_1$ as follows: If the set $\{y_1, y_2, ..., y_n\}$ and the set $\mathfrak{I}_1$ have a term in common, say h_k, then replace h_k with xh_k in the set $\mathfrak{I}_1$. That is:

$$\text{If } h_k \in \mathfrak{I}_1 \cap \{y_1, y_2, ..., y_n\} \text{ then } h_k \leftarrow x h_k \tag{15.4}$$

Repeat Eq. 15-4 for all k and as many times as needed until $\{y_1, y_2, ..., y_n\}$ and $\mathfrak{I}_1$ no longer have any terms in common. This may require some terms h_k to be multiplied by x multiple times.

3.) Once $\mathfrak{I}_1 \cap \{y_1, y_2, ..., y_n\} = \varnothing$, then form a new set $\mathfrak{I}_2$ composed of all of the elements of $\mathfrak{I}_1$ and all of their derivatives:

$$\mathfrak{I}_2 = \{h_1, h_1', h_1'', ..., h_2, h_2', h_2'', ...\} \tag{15.5}$$

4.) Take out terms in common with $\{y_1, y_2, ..., y_n\}$:

$$\mathfrak{I} = \mathfrak{I}_2 \setminus (\mathfrak{I}_2 \cap \{y_1, y_2, ..., y_n\}) \tag{15.6}$$

5.) Assume the particular solution is in the form of all linear combinations of the elements of $\mathfrak{I} = \{f_1, f_2, ...\}$:

$$y_p = A_1 f_1 + A_2 f_2 + ... + A_m f_m \tag{15.7}$$

Plug y_p into the non-homogeneous equation and work out what the A's are.

Example 15-1: Solve the constant coefficient non-homogeneous equation:

$$y'' + 3y' - 10y = 6e^{4x} \qquad (15.8)$$

Solution: The first step is to solve the homogeneous equation:

$$y'' + 3y' - 10y = 0 \qquad (15.9)$$

Letting $y = e^{mx}$, we find the characteristic equation:

$$m^2 + 3m - 10 = 0 \quad \Rightarrow \quad (m - 2)(m + 5) = 0 \qquad (15.10)$$

This has two solutions: $\{y_1, y_2\} = \{e^{2x}, e^{-5x}\}$.

Next we take a hard look at g(x) and see if we can tell what summands it has. The only one we have is $g_1 = e^{4x}$. Thus:

$$\mathfrak{I}_1 = \{e^{4x}\} \qquad (15.11)$$

Since $\{y_1, y_2\}$ and $\mathfrak{I}_1$ have no terms in common, we don't have to multiply anything in $\mathfrak{I}_1$ by x. Thus $\mathfrak{I}_2$ is also $\{e^{4x}\}$:

$$\mathfrak{I}_2 = \{e^{4x}\} \qquad (15.12)$$

Now the derivative of e^{4x} is just e^{4x} again, scaled by the constant 4, so $\mathfrak{I}$ is the same as $\mathfrak{I}_2$:

$$\mathfrak{I} = \{e^{4x}\} \qquad (15.13)$$

So we assume particular solution of the form:

$$y_p = Ae^{4x} \qquad (15.14)$$

Then, making use of this by plugging in, we have:

$$16Ae^{4x} + 12Ae^{4x} - 10Ae^{4x} = 6e^{4x} \qquad 15.15)$$

This tells us that A = 1/3. Thus $y_p = (1/3)e^{4x}$ and the general solution to the non-homogeneous equation is:

$$y = C_1 e^{-5x} + C_2 e^{2x} + \frac{1}{3} e^{4x} \qquad (15.16)$$

◊

Example 15-2: Solve:

$$y'' + y = 2\cos x \qquad (15.17)$$

Solution: The set of orthogonal functions that solve the homogeneous equation $y'' + y = 0$ are $\{y_1, y_2\} = \{\sin x, \cos x\}$. The non-homogeneous term $g(x)$ has only one summand, $g_1 = \cos x$; therefore, we start with the following function set:

$$\mathfrak{I}_1 = \{\cos(x)\} \qquad (15.18)$$

This $\mathfrak{I}_1$ has a term in common with $\{y_1, y_2\}$; therefore, we multiply it by x and replace it in order to form $\mathfrak{I}_2$. First:

$$\mathfrak{I}_1 = \{x\cos(x)\} \qquad (15.19)$$

We now have $\mathfrak{I}_1 \cap \{y_1, y_2\} = \varnothing$. So the next step is to augment $\mathfrak{I}_1$ with all of its derivatives in order to form $\mathfrak{I}_2$:

$$\mathfrak{I}_2 = \{-x\sin(x), x\cos(x), \cos(x)\} \qquad (15.20)$$

Again, subtracting off the intersection with $\{y_1, y_2\}$, and ignoring a minus sign, we have:

$$\mathfrak{I} = \{x\sin(x), x\cos(x)\} \qquad (15.21)$$

Thus we can assume y_p is in the form:

$$y_p = A_1 x\sin(x) + A_2 x\cos(x) \qquad (15.22)$$

Plugging this into Eq. 15-17, we obtain:

$$\left(2A_1\cos x - A_1 x\sin x - 2A_2\sin x - A_2 x\cos x\right) + $$
$$A_1 x\sin(x) + A_2 x\cos(x) = 2\cos x \qquad (15.23)$$

Or, after some canceling:

$$2A_1 \cos x - 2A_2 \sin x = 2\cos x \tag{15.24}$$

Matching like terms, we see that $A_1 = 1$ and $A_2 = 0$. The particular solution then is:

$$y_p = x\sin x \tag{15.25}$$

Thus the solution to Eq. 15-17 is:

$$y = C_1 \cos x + C_2 \sin x + x\sin x \tag{15.26}$$

◊

EXERCISES

Exercise 15-1: Find the general solution to $y'' + 4y = 3\sin(x)$.

Exercise 15-2: Find the general solution to $2y'' + 3y' + y = \cos(x)$.

Exercise 15-3: Find the general solution to $y'' - 2y' - 3y = 2e^{3x} + 8xe^{3x}$.

Article 16. Variation of parameters

Wronski was born in Poland but spent most of his life in France. We have been introduced to the Wronskian, the determinant

$$W = \begin{vmatrix} y_1 & y_2 \\ y_1' & y_2' \end{vmatrix}$$

(16.1)

Wronski showed that, if y_1 and y_2 (on some interval I) are two solutions of the second-order linear homogeneous ODE, then they are linearly independent on I if $W \neq 0$ for at least one point $x \in I$.

Hoëné Wronski
(1778-1853)

One could claim that, even if the Wronskian determinant was his only contribution to mathematics, Wronski's name would still be remembered today...a claim that is fortified by the fact that the Wronskian determinant actually *was* Wronski's only contribution to mathematics.

This article presents another method for finding a particular solution y_p for a non-homogeneous linear ODE. Again, the reason we need to determine a particular solution y_p is so that we can add it on to the homogeneous solution y_h = $C_1 y_1 + C_2 y_2$ in order to form the general solution:

$$y_g = C_1 y_1 + C_2 y_2 + y_p$$

(16.2)

However, unlike the method of the previous article, the present **variation of parameters method** gives a means for finding a particular solution when we have *non-constant coefficients*:

$$y'' + P(x)y' + Q(x)y = g(x)$$

(16.3)

Note that we have provided for the coefficient in front of y'' to be one (by having divided through by whatever was there when the problem was initially posed). For using variation of parameters, it will be important to start out in this form.

To use the variation of parameters technique, we begin by solving the homogeneous equation $y'' + P(x)y' + Q(x)y = 0$. The homogeneous solution then is:

$$y_h = C_1 y_1 + C_2 y_2 \qquad (16.4)$$

The basis functions $\{y_1, y_2\}$ of the solution space of the homogeneous equation will be useful in deciding what the particular solution y_p can be.

Next we make the following assumption for the form of y_p:

$$y_p = u_1(x)y_1(x) + u_2(x)y_2(x) \qquad (16.5)$$

Here we have introduced new *functional coefficients* $u_1(x)$ and $u_2(x)$. Eq. 16-3 might look like it "fell out of the sky" but it's actually straightforward to see why it makes sense. We can see that assuming functional coefficients in front of the y's is correct by looking at what is absurd about assuming constant coefficients in front of the y's instead.

If we assume (wrongly) $y_p = A_1 y_1 + A_2 y_2$ with A_1 and A_2 constant, then the left side of $y'' + P(x)y' + Q(x)y = g(x)$ becomes zero. (This is because (A_1, A_2) is a point in the solution space of $y'' + P(x)y' + Q(x)y = 0$.) Stop and think about this for a moment. Thus the constant coefficient assumption forces $g(x)$ to zero which contradicts our initial assumption of having a non-homogeneous equation and $g(x) \neq 0$. Thus we cannot have a particular solution in the form $y_p = A_1 y_1 + A_2 y_2$ with A_1 and A_2 constant; rather, we have to assume functional coefficients $u_1(x)$ and $u_2(x)$ when we express y_p in terms of y_1 and y_2, that is, we must assume $y_p = u_1(x)y_1(x) + u_2(x)y_2(x)$.

The implications of the assumption that y_p can be written as the sum of constants times the y's are as follows:

$$y_p = A_1 y_1 + A_2 y_2 \quad \overset{IMPLIES}{\Rightarrow}$$

$$\frac{d^2}{dx^2}(A_1 y_1 + A_2 y_2) + P(x)\frac{d}{dx}(A_1 y_1 + A_2 y_2) +$$

$$Q(x)(A_1 y_1 + A_2 y_2) = g(x) \quad \overset{IMPLIES}{\Rightarrow}$$

$$A_1 \left(\frac{d^2 y_1}{dx^2} + P(x)\frac{dy_1}{dx} + Q(x)y_1 \right) +$$

$$A_2 \left(\frac{d^2 y_2}{dx^2} + P(x)\frac{dy_2}{dx} + Q(x)y_2 \right) = g(x) \quad \overset{IMPLIES}{\Rightarrow}$$

$$A_1(0) + A_2(0) = g(x) \quad \overset{IMPLIES}{\Rightarrow} \quad g(x) = 0 \tag{16.6}$$

This is like saying "a non-homogeneous equation always has the homogeneous term equal to zero" …an idea that is worthy of our contempt! Instead the form we should assume is:

$$y_p = u_1(x)y_1(x) + u_2(x)y_2(x) \tag{16.7}$$

The above equation is the heart of variation of parameters, a method introduced by Joseph-Louis Lagrange. If we plug Eq. 16-7 into $y'' + Py' + Qy = g(x)$, we get:

$$u_1'' y_1 + 2u_1' y_1' + u_1 y_1'' + u_2'' y_2 + 2u_2' y_2' + u_2 y_2'' + \\ Pu_1' y_1 + Pu_1 y_1' + Pu_2' y_2 + Pu_2 y_2' + Qu_1 y_1 + Qu_2 y_2 = g(x) \tag{16.8}$$

Or:

$$u_1\{y_1'' + Py_1' + Qy_1\} + u_2\{y_2'' + Py_2' + Qy_2\} + P\{u_1' y_1 + u_2' y_2\} + \\ \frac{d}{dx}\{u_1' y_1 + u_2' y_2\} + u_1' y_1' + u_2' y_2' = g(x) \tag{16.9}$$

Since y_1 and y_2 are homogeneous solutions, the first two terms reduce to zero:

$$u_1\{0\} + u_2\{0\} + P\{u_1'y_1 + u_2'y_2\} + \frac{d}{dx}\{u_1'y_1 + u_2'y_2\} + u_1'y_1' + u_2'y_2' =$$

$$g(x)$$

(16.10)

Let's look carefully at what remains:

$$P\{u_1'y_1 + u_2'y_2\} + (^d/_{dx})\{u_1'y_1 + u_2'y_2\} + u_1'y_1' + u_2'y_2' = g(x) \qquad (16.11)$$

If we let the term in bold equal zero then we have two equations in two unknowns for solving for the u's:

$$u_1'y_1 + u_2'y_2 = 0$$
$$u_1'y_1' + u_2'y_2' = g(x)$$

(16.12)

We are justified in setting $u_1'y_1 + u_2'y_2 = 0$ for the reason that Eq. 16-5 left us enough "wiggle room" in order to do so. Eq. 16-5 represents one equation in the two unknowns u_1 and u_2. But we need to have two equations in the two unknowns in order to determine u_1 and u_2 and thus y_p. Introducing $u_1'y_1 + u_2'y_2 = 0$ gives us the needed second equation, consistent with y_p being a solution to the non-homogeneous equation.

Using Eq. 16-12, we can now solve for the derivatives of u_1 and u_2. We may use the first equation to solve for u_2':

$$u_2' = -u_1' \frac{y_1}{y_2}$$

(16.13)

Plugging this into the second equation, we get:

$$u_1' = -\frac{y_2 g}{(y_1 y_2' - y_1' y_2)}$$

(16.14)

But the denominator is just the Wronskian; therefore,

$$u_1' = -\frac{y_2 g}{W}$$

(16.15)

Plugging Eq. 16-15 into Eq. 16-13, we can solve for u_2':

$$u_2' = \frac{y_1 g}{W}$$

(16.16)

So an algorithm for using variation of parameters is as follows:

<u>Variation of parameters algorithm:</u>

1.) Provide for the differential equation to be in the form $y'' + P(x)y' + Q(x)y = g(x)$.
2.) Find y_1 and y_2, the solutions to the homogeneous equation $y'' + P(x)y' + Q(x)y = 0$.
3.) Calculate W and u_1' and u_2'.
4.) Integrate u_1' and u_2' to form u_1 and u_2. The particular solution is $y_p = u_1 y_1 + u_2 y_2$.

With particular solution in hand, the general solution then is just:

$$y_g = y_h + y_p = C_1 y_1 + C_2 y_2 + u_1 y_1 + u_2 y_2$$

(16.17)

Example 16-1: Solve:

$$y'' - 2y' + y = 2x$$

(16.18)

The equation is already in standard form. So the next step is to solve the homogeneous equation:

$$m^2 e^{mx} - 2me^{mx} + e^{mx} = 0$$

(16.19)

This reduces to the characteristic equation $(m-1)(m-1) = 0$ which gives the double root $m = 1$. Thus we have $y_1 = e^x$. For a second solution to the homogeneous equation, let $y_2 = v(x)y_1 = v(x)e^x$. This gives us a second homogeneous solution $y_2 = xe^x$. So the Wronskian is:

$$W = \begin{vmatrix} e^x & xe^x \\ e^x & e^x(x+1) \end{vmatrix} = e^{2x}(x+1) - xe^{2x} = e^{2x}$$

(16.20)

We can solve for the derivatives of the u's:

$$u_1' = -\frac{y_2 g}{W} = -\frac{xe^x 2x}{e^{2x}} = -2x^2 e^{-x}$$

$$u_2' = \frac{y_1 g}{W} = \frac{e^x 2x}{e^{2x}} = 2xe^{-x} \tag{16.21}$$

Now we can integrate and solve for u_1:

$$u_1 = -2\int x^2 e^x \, dx = 2x^2 e^{-x} + 4xe^{-x} + 4xe^{-x} \tag{16.22}$$

We can get u_2 just as easily:

$$u_2 = 2\int xe^{-x} \, dx = 2\left[(-1)\left(xe^{-x} - \int e^{-x} \, dx\right)\right] = -2xe^{-x} - 2e^{-x} \tag{16.23}$$

Now $y_p = u_1 y_1 + u_2 y_2$ and plugging in we get:

$$y_p = 2x + 4 \tag{16.24}$$

So the general solution is:

$$y = C_1 e^x + C_2 xe^x + 2x + 4 \tag{16.25}$$

◊

EXERCISES

Exercise 16-1: Find a particular solution to:

$$y'' + 9y = \tan 3x \tag{16.26}$$

Exercise 16-2: Use variation of parameters to solve:

$$y'' - y = e^{2x} \tag{16.27}$$

Exercise 16-3: Find the general solution of:

$$y'' + 2y' + y = e^{3x} \qquad (16.28)$$

Exercise 16-4: Find the general solution of:

$$y'' + y = \csc(x) \qquad (16.29)$$

Article 17. Cauchy-Euler Equation

Notice that, in the equation below, the power to which x is raised in each term is the same as the order of derivative of y in that term:

$$a_n x^n \frac{d^n y}{dx^n} + a_{n-1} x^{n-1} \frac{d^{n-1} y}{dx^{n-1}} + \ldots + a_1 x \frac{dy}{dx} + a_0 y = f(x)$$

(17.1)

An equation in the form of Eq. 17-1 is known as a Cauchy-Euler equation. An approach for finding homogeneous solutions (the $f(x) = 0$ case) is to look for solutions in the form:

$$y = x^m$$

(17.2)

Let's consider the 2nd-order homogeneous Cauchy-Euler equation:

$$a x^2 \frac{d^2 y}{dx^2} + b x \frac{dy}{dx} + c y = 0$$

(17.3)

Plugging in $y = x^m$, we have:

$$a x^2 m(m-1)x^{m-2} + b \, xmx^{m-1} + c x^m = 0$$

(17.4)

This becomes:

$$a \, m(m-1)x^m + b \, mx^m + c x^m = 0$$

(17.5)

Dividing out x^m leaves an auxiliary equation:

$$am^2 + (b - a)m + c = 0$$

(17.6)

How we make use of the m's will give you a sense of déjà vu. For the Cauchy-Euler equation, there are three possible forms the homogeneous solution y_h can take depending on whether the roots of Eq. 17-6 are 1.) distinct real m_1 and m_2, 2.) repeated real m, or 3.) complex conjugate roots $m = \alpha \pm i \beta$.

Table 17-1: Homogeneous solution table for 2nd-order Cauchy-Euler equation.

Case 1	Distinct real roots	$y_h = C_1 x^{m_1} + C_2 x^{m_2}$
Case 2	Repeated real root	$y_h = C_1 x^m + C_2 x^m \ln x$
Case 3	Complex roots	$y_h = x^\alpha [C_1 \cos(\beta \ln x) + C_2 \sin(\beta \ln x)]$

Example 17-1: Solve $x^2 y'' + xy' - 9y = 0$.

Solution: Substituting $y = x^m$ we have:

$$x^2 m(m-1)x^{m-2} + xmx^{m-1} - 9x^m = 0 \qquad (17.7)$$

Or:

$$m(m-1)x^m + mx^m - 9x^m = 0 \qquad (17.8)$$

Dividing out x^m provides us the auxiliary equation:

$$m(m-1) + m - 9 = 0 \quad \Rightarrow \quad m^2 - 9 = 0 \qquad (17.9)$$

Thus we have two distinct real roots, $m_1 = 3$, $m_2 = -3$. So, referring to table 17-1, we see that the solution is:

$$y = C_1 e^{3x} + C_2 e^{-3x} \qquad (17.10)$$

◊

But where does Table 17-1 come from? We can derive Table 17-1 by making the substitution $z = ln(x)$ in Table 13-1. (Recall that Article 13 was concerned with finding solutions to equations with constant coefficients.) First replace x with z in Table 13-1.

Table: Constant coefficients homogeneous solutions.

Case 1	Distinct real roots	$y_h = C_1 e^{m_1 z} + C_2 e^{m_2 z}$
Case 2	Repeated real root	$y_h = C_1 e^{mz} + C_2 x e^{mz}$
Case 3	Complex roots	$y_h = e^{\alpha z}\left[C_1 \cos(\beta z) + C_2 \sin(\beta z)\right]$

Next introduce $z = ln(x)$:

Case 1	Distinct real roots	$y_h = C_1 e^{m_1 \ln x} + C_2 e^{m_2 \ln x}$
Case 2	Repeated real root	$y_h = C_1 e^{m \ln x} + C_2 x e^{m \ln x}$
Case 3	Complex roots	$y_h = e^{\alpha \ln x}\left[C_1 \cos(\beta \ln x) + C_2 \sin(\beta \ln x)\right]$

This reduces to Table 17-1 by using some properties of logarithms:

$$e^{m \ln x} = e^{\ln x^m} = x^m$$

$$e^{\alpha \ln x} = e^{\ln x^{\alpha}} = x^{\alpha}$$

$$(17.11)$$

How is it that we can derive the solution table for solutions to the Cauchy-Euler equation so easily from the solution table for the constant coefficients equation? The answer is that $x = e^z$ turns the Cauchy-Euler equation into a constant coefficients equation:

$$a\frac{d^2 y}{dz^2} + (b-a)\frac{dy}{dz} + c y = 0$$

$$(17.12)$$

The Cauchy-Euler equation is "a change of variables away" from being a constant coefficient equation, where the change of variables is $x = e^z$ or $z = \ln(x)$. Remember the change of variables is EASY. (Get it? e to the z.)

Example 17-2: Show that the Cauchy-Euler equation $ax^2 y'' + bxy' + cy = 0$ reduces to a 2nd-order linear ODE with constant coefficients by making the substitution $x = e^z$.

Solution: Making the substitution $x = e^z$, the derivatives with respect to x can be converted as follows:

$$\frac{dy}{dx} = \frac{dy}{dz}\frac{dz}{dx} = \frac{dy}{dz}e^{-z} \quad \Rightarrow \quad \frac{d^2y}{dx^2} = e^{-2z}\frac{d^2y}{dz^2} - e^{-2z}\frac{dy}{dz}$$
(17.13)

Plugging $x = e^z$ and these derivatives into $ax^2y'' + bxy' + cy = 0$, we get:

$$ae^{2z}\left(e^{-2z}\frac{d^2y}{dz^2} - e^{-2z}\frac{dy}{dz}\right) + be^z\left(\frac{dy}{dz}e^{-z}\right) + cy = 0$$
(17.14)

Which reduces to:

$$a\frac{d^2y}{dz^2} + (b-a)\frac{dy}{dz} + cy = 0$$
(17.15)

Which is a constant coefficient equation.

◊

Let's do another example using Table 17-1.

Example 17-3: Solve:

$$4x^2y'' + y = 0$$
(17.16)

Solution: Substituting $y = x^m$, we have:
$$4x^2m(m-1)x^{m-2} + x^m = 0 \quad \Rightarrow \quad 4m(m-1)x^m + x^m = 0$$
(17.17)

Whose auxiliary equation is:

$$4m(m-1) + 1 = 0 \quad \Rightarrow \quad m^2 - m + \frac{1}{4} = 0$$
(17.18)

Which factors as $(m - \frac{1}{2})^2 = 0$. Thus $m = \frac{1}{2}$ is a double root; therefore, referring to Table 17-1, the solution is:

$$y = C_1\sqrt{x} + C_2\sqrt{x}\,(\ln x) \tag{17.19}$$

◊

Augustin Cauchy
(1789-1857)

Cauchy was born a few years after Euler died, a factor that must have made two-way collaboration on the Cauchy-Euler equation challenging. ☺ Cauchy invented the determinant and contributed enormously to complex analysis as well as to many other areas of mathematics. Let's solve a Cauchy-Euler equation of the third case, one having an auxiliary equation with complex conjugate roots.

Example 17-4: Solve $x^2 y'' + xy' + 4y = 0$.

Solution: Plugging in $y = x^m$:

$$x^2 m(m-1)x^{m-2} + xmx^{m-1} + 4x^m = 0$$
$$\Downarrow$$
$$m(m-1)x^m + mx^m + 4x^m = 0 \tag{17.20}$$

Dividing out x^m and simplifying, we have the auxiliary equation:

$$m^2 + 4 = 0 \tag{17.21}$$

This has complex conjugate roots $m = \pm i$. Referring back to Table 17-1, we see that the solution is:

$$y = x^0 \left[C_1 \cos(2\ln x) + C_2 \sin(2\ln x) \right] \tag{17.22}$$

Or:

113

$$y = C_1 \cos(2\ln x) + C_2 \sin(2\ln x)$$

(17.23)

◊

The examples above are for the homogeneous case $f(x) = 0$. For problems where $f(x) \neq 0$ we would proceed as before: First find the homogeneous solution $y_h = C_1y_1 + C_2y_2$ and then find a particular solution y_p. The general solution then is the sum of these: $y_g = y_h + y_p$.

_____

EXERCISES

Exercise 17-1: Find the general (homogeneous) solution to $x^2y'' + 13xy' + 20y = 0$.

Exercise 17-2: Find the general solution to the non-homogeneous equation:

$$x^2y'' + 3xy' + y = \pi$$

(17.24)

Exercise 17-3: Find the general solution of $4x^2y'' + 8xy' + 5y = 0$.

_____

Article 18. Series Solution

The method of solving differential equations by series solution is about as "brute force" as you can get. The algorithm goes as follows: You simply write down a series,

$$y = \sum_{n=0}^{\infty} c_n x^n \qquad (18.1)$$

Where the c's are constants and x is the independent variable, and then plug the series into the differential equation at hand. After reducing the result, you determine what formula for generating the c's becomes apparent. Then you also need to establish what interval of convergence--if any--this series has.

The series solution approach can become very tedious to do by hand. The benefit though is that it can be used against very general types of cases such as non-constant coefficient cases. Let's turn to a very easy example to begin to get a feel for how the method works. In the analysis that follows, we will need to keep in mind that the derivative dy/dx is a linear operator:

$$\frac{d}{dx}(a_1 f_1(x) + a_2 f_2(x)) = a_1 \frac{d}{dx}(f_1(x)) + a_2 \frac{d}{dx}(f_2(x)) \qquad (18.2)$$

This will allow us to bring the derivative into a series by allowing it to "pass through" the summation:

$$\frac{d}{dx}\left(\sum_{n=0}^{\infty} c_n x^n \right) = \sum_{n=0}^{\infty} c_n \frac{d}{dx}\left(x^n \right) \qquad (18.3)$$

Example 18-1: Solve $xy + y' = 0$.

Plugging in the series we get:

$$(x)\sum_{n=0}^{\infty} c_n x^n + \frac{d}{dx}\left(\sum_{n=0}^{\infty} c_n x^n \right) = 0 \qquad (18.4)$$

For the leftmost term, we can use the distributive law to bring the multiplication by x into the series. For the second term, the derivative can be brought into the summation since it is a linear operator.

115

$$\sum_{n=0}^{\infty} c_n x^{n+1} + \sum_{n=0}^{\infty} c_n \frac{d}{dx}\left(x^n\right) = 0 \qquad (18.5)$$

Which equals:

$$\sum_{n=0}^{\infty} c_n x^{n+1} + \sum_{n=0}^{\infty} c_n n x^{n-1} = 0 \qquad (18.6)$$

If we can add the two terms appearing on the left, we can arrive at a formula that generates the coefficients. The problem is that the powers of x don't match. Let us write out a few terms of Eq. 18-6 to see what's going on:

$$(c_0 x + c_1 x^2 + \ldots) + (c_0(0) + c_1 + 2c_2 x^1 + 3c_3 x^2 + \ldots) = 0 \qquad (18.7)$$

But this is the same as:

$$(c_0 x + c_1 x^2 + \ldots) + (c_1 + 2c_2 x + 3c_3 x^2 + \ldots) = 0 \qquad (18.8)$$

The subscript on the constants on the left lag by two steps those on the right, from the point of view of desiring to have the powers of x correspond. We can rewrite the second series in Eq. 18-6 so that the powers of x correspond in a way that is conducive to combining the series by "advancing subscripts" in the second series. Note carefully how we rewrite the above--especially the second term. Looking at the second term in Eq. 18-8, we see that replacing n with n+2 and adding the c_1 term "out front" creates exactly the same series as Eq. 18-6 except that now the powers of x are compatible:

$$\sum_{n=0}^{\infty} c_n x^{n+1} + \left(c_1 + \sum_{n=0}^{\infty} (n+2)c_{n+2} x^{n+1} \right) = 0 \qquad (18.9)$$

We are justified to "bring out" c_1 from the right entry since there is no x-to-the-zero-power term on the left side to match it with. Now we can add the series:

$$\sum_{n=0}^{\infty} \left(c_n + (n+2)c_{n+2} \right) x^{n+1} + c_1 = 0 \qquad (18.10)$$

Now, zero, on the right-hand side, is a polynomial (with zero coefficients). Equal polynomials are equal term by term; therefore,

116

$$c_1 = 0 \quad and \quad c_n + (n+2)c_{n+2} = 0 \quad \Rightarrow \quad (n+2)c_{n+2} = -c_n \quad (18.11)$$

So we have a **recursion formula** which generates the coefficients:

$$c_{n+2} = -\frac{c_n}{n+2} \qquad (18.12)$$

Now all of the odd coefficients have to be zero; because if c_1 is zero then c_3 is zero, etc. The even coefficients are a different story. If we start with c_0 we have:

$$c_2 = -\frac{c_0}{2}$$

$$c_4 = -\frac{c_2}{4} = -\frac{1}{4}\left(-\frac{c_0}{2}\right) = \frac{1}{4\times2}c_0 \qquad (18.13)$$

$$c_6 = -\frac{1}{6\times4\times2}c_0$$

Etc.

We recognize a pattern:

$$c_{2n} = (-1)^n \frac{2^{-n}}{n!} \qquad (18.14)$$

Thus the solution is:

$$y = \sum_{n=0}^{\infty} c_n x^n \qquad (18.15)$$

Or:

$$y = c_0 \sum_{n=0}^{\infty} (-1)^n \frac{2^{-n}}{n!} x^{2n} \qquad (18.16)$$

Or:

$$y = c_0 \sum_{n=0}^{\infty} \frac{1}{n!} \left(-\frac{x^2}{2} \right)^n \tag{18.17}$$

But the Maclaurin expansion for the exponential is:

$$\sum_{n=0}^{\infty} \frac{1}{n!} \varphi^n = e^{\varphi} \tag{18.18}$$

So $\varphi = -\frac{1}{2} x^2$ and the final answer is:

$$y = c_0 e^{-x^2/2} \tag{18.19}$$

Which converges for all x.

◊

This example suggests a rough algorithm for applying series solution:
Algorithm for Series Solution:

1.) Plug in the series $y = \sum_{n=0}^{\infty} c_n x^n$.

2.) Distribute coefficients and derivatives so that they appear within the summations.

3.) Advance indices as needed so that powers of x correspond if multiple summations appear. (It may be helpful to write out a few terms of the summations to understand how to perform this. And this step may result in constant terms appearing as separate summands which do not participate in any of the "sigma summations.")

4.) Combine multiple summations by adding like terms.

5.) Determine coefficient generating function(s) by matching like terms on the left and right sides of the equation.

6.) Try to reduce the coefficient generating functions to recursion relations.

7.) Try to recognize any closed form functions that the recursion relations may correspond to.

Let's apply our algorithm to the solution of an equation which is extremely important to engineering applications. Bessel's equation of order p, where p is a non-negative constant is:

$$x^2 y'' + x y' + \left(x^2 - p^2 \right) y = 0 \tag{18.20}$$

Let's restrict our attention to the zeroeth-order case:

$$x^2 y'' + x y' + \left(x^2 - p^2\right) y = 0 \tag{18.21}$$

$$x^2 y'' + x y' + x^2 y = 0 \tag{18.22}$$

Plugging in the series substitution we obtain:

$$x^2 \sum_{n=0}^{\infty} n(n-1) c_n x^{n-2} + x \sum_{n=0}^{\infty} n c_n x^{n-1} + x^2 \sum_{n=0}^{\infty} c_n x^n = 0 \tag{18.23}$$

Bringing the coefficient functions into the summations, we have:

$$\sum_{n=0}^{\infty} n(n-1) c_n x^n + \sum_{n=0}^{\infty} n c_n x^n + \sum_{n=0}^{\infty} c_n x^{n+2} = 0 \tag{18.24}$$

Writing out a few terms, we see that:

$$\left(2 c_2 x^2 + 3 \times 2 c_3 x^3 + \ldots\right) + \left(c_1 x + 2 c_2 x^2 + \ldots\right) + \left(c_0 x^2 + c_1 x^3 + \ldots\right) = 0 \tag{18.25}$$

If we advance exponents in the left two summands in Eq. 18-24, we have, being careful not to discard the $c_1 x$ term:

$$\sum_{n=0}^{\infty} (n+2)(n+1) c_{n+2} x^{n+2} + c_1 x +$$

$$\sum_{n=0}^{\infty} (n+2) c_{n+2} x^{n+2} + \sum_{n=0}^{\infty} c_n x^{n+2} = 0 \tag{18.26}$$

These series can now be added:

$$\sum_{n=0}^{\infty} \left\{(n+2)(n+1) c_{n+2} + (n+2) c_{n+2} + c_n\right\} x^{n+2} + c_1 x = 0 \tag{18.27}$$

Or:

119

$$\sum_{n=0}^{\infty} \left\{ (n+2)^2 c_{n+2} + c_n \right\} x^{n+2} + c_1 x = 0$$

$$(18.28)$$

On the right-hand side of Eq. 18-28 we have the polynomial with all coefficients zero. Thus, comparing like terms on the left and right of Eq. 18-28, we have:

$$c_1 = 0$$
$$(n+2)^2 c_{n+2} + c_n = 0$$

$$(18.29)$$

This gives the coefficient generator:

$$c_1 = 0$$
$$c_0 = c_0$$
$$c_{n+2} = \frac{-c_n}{(n+2)^2}$$

$$(18.30)$$

Which generates coefficients for the even-powered terms only. (All of the odd-powered terms are zero since $c_1 = 0$.) If we use this generating function to produce a few terms, we begin to recognize a pattern:

$$c_0 = c_0 = \frac{c_0}{(0!)^2 (2^2)^0}$$

$$c_2 = \frac{-c_0}{2^2} = (-1)\frac{c_0}{(2\times1)^2} = (-1)\frac{c_0}{(1!)^2 (2^2)^1}$$

$$c_4 = \frac{-c_2}{4^2} = \frac{c_0}{4^2 2^2} = \frac{c_0}{(2\times2)^2 (2\times1)^2} = \frac{c_0}{(2!)^2 (2^2)^2}$$

$$c_6 = \frac{-c_4}{6^2} = \frac{-c_0}{6^2 4^2 2^2} = (-1)\frac{c_0}{(3\times2)^2 (2\times2)^2 (2\times1)^2} =$$

$$(-1)\frac{c_0}{(3!)^2 (2^2)^3}$$

...

$$(18.31)$$

$$c_{2n} = (-1)^n \frac{c_0}{(n!)^2 2^{2n}}$$

(18.32)

So the series solution then is:

$$y = c_0 \sum_{n=0}^{\infty} (-1)^n \frac{1}{(n!)^2 2^{2n}} x^{2n}$$

(18.33)

◊

Functions such as this one that satisfy Bessel's equation are called **Bessel functions**. For applications, much of the interest in differential equations is focused on the functions they produce. Equations like Bessel's equation produce families of functions that can be used to represent engineering functions and signals in a way that greatly reduces the difficulty of engineering problems. For example, problems involving propagation of waves from a source having cylindrical symmetry are often greatly simplified when expressed using Bessel functions.

_____

Exercise 18-1: Use the Series Solution to find the general solution to $y'' + 49y = 0$.

Exercises 18-2 through 18-4 refer to Hermite's equation:

$$y'' - 2xy' + 2py = 0$$

(18.34)

The first three Hermite functions are 1, x, and $1 - 2x^2$.

Exercise 18-2: Find a solution of the $p = 0$ case of Hermite's equation.

Exercise 18-3: Find a solution of the $p = 1$ case of Hermite's equation.

Exercise 18-4: Find a solution of the $p = 2$ case of Hermite's equation.

Part three:
Laplace Transforms

Article 19. Introduction to the Laplace Transform

For the rest of this text, we will be discussing **transform methods** for solving problems involving differential equations. Up to this point, the kinds of methods we have been using have been *in situ* techniques. "In situ" is Latin meaning "in place." For the most part, the problems we have seen so far have been solved *in place* by plugging in substitutions which result in simplifications that, in turn, eventually result in solution. For the transform methods that we take up presently, our philosophy will be quite different: Rather than solve the problem *in situ*, we will *transform* the problem into a new domain-- a realm in which the problem is more tractable--and work on it in that new domain. Then once the problem is solved in the transform domain, we will transform the answer back to the original realm.

Once we solve the problems in the transform domain, we will **inverse transform** (also called reverse transform) the solutions back to the original problem domain and report the answer.

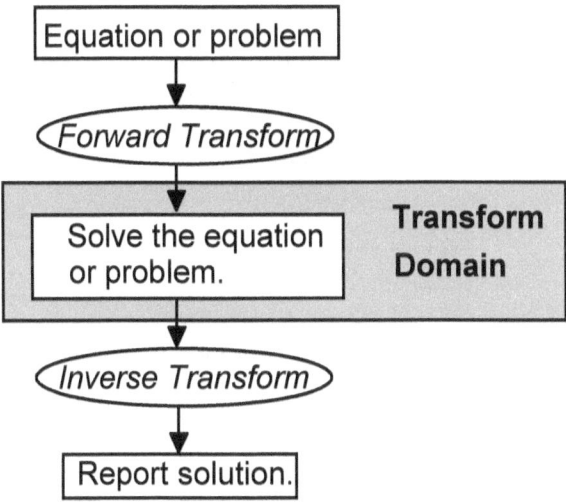

Figure 19-1: Transform Approach to Problem Solving.

The **Laplace transform** will be the particular type of transform that we consider. We will take problems involving differential equations (that would be hard to solve *in situ*) and transform them into the Laplace domain and find that they are easy to tackle in the Laplace domain. Once we have solved the problems in the Laplace domain, we will inverse-Laplace transform these problems back to the original setting of the problem.

Shortcut to Ordinary Differential Equations

We turn now to the definition of a Laplace transform:

Definition 19-1: The **Laplace transform** of a function f(t) is the inner product between f(t) and an exponential function of the form e^{-st}:

$$\mathcal{L}\big(f(t)\big) \;=\; \int_0^\infty f(t)e^{-st}dt$$

(19.1)

Note that the integration is from zero to infinity. The result of this integration is to produce a function of *s*, that is; *t* is only a dummy variable; therefore, *t* is only used to accomplish the integration and then is discarded. The original function is a function of *t*, in particular *f(t)*; however, in the transform domain we have a function of *s*. This rendering of *f* in the transform domain is written *F(s)*.

$$F(s) \;=\; \int_0^\infty f(t)e^{-st}dt$$

(19.2)

The capitol letter indicates a function in the transform domain. Therefore, y(t) in the Laplace transform domain becomes Y(s), x(t) becomes X(s), and so on.

The inverse Laplace transform is as follows:

$$f(t) \;=\; \frac{1}{2\pi i}\int_{c-i\infty}^{c+i\infty} F(s)e^{ts}ds$$

(19.3)

Which is true for c "large enough." It is rare for this latter integral, discovered by Poisson, to be used directly.

A function in the input domain and its counter-part in the transform domain are referred to as transform pairs. The inverse transform is usually never calculated because Eq. 19-2 is sufficient to generate the needed *transform pairs*. That is, all of the means for connecting a function to its Laplace domain counterpart is provided by Eq. 19-2.

Let's now take a moment to see what some familiar functions become when they are transformed into the Laplace domain. First, let's determine the Laplace transform of unity.

126

$$\mathcal{L}(1) \;=\; \int_0^\infty 1\, e^{-st} dt \qquad \Rightarrow$$

$$-\frac{1}{s}\int_0^\infty e^{-st}\left(-s\right)dt \;=\; -\frac{1}{s}e^{-st}\,\Big|_0^\infty \;=\; -\frac{1}{s}(0-1) \;=\; \frac{1}{s}$$

$$(19.4)$$

So y(t) = 1 in the transform domain becomes Y(s) = 1/s in the Laplace domain. The Laplace transform of t is also not difficult.

$$\mathcal{L}(t) \;=\; \int_0^\infty t\, e^{-st} dt \qquad \Rightarrow \qquad -\frac{1}{s}\int_0^\infty t\, e^{-st}\left(-s\right)dt$$

$$(19.5)$$

We can use integration by parts as follows:

$$\text{Let } u = t \;\text{ and }\; dv = e^{-st}\left(-s\right)dt$$

$$\mathcal{L}(t) \;=\; -\frac{1}{s}e^{-st}\, t \,\Big|_0^\infty \;-\; \left(-\frac{1}{s}\int_0^\infty e^{-st}\, dt\right)$$

$$=\; 0 \;+\; \frac{1}{s}\int_0^\infty e^{-st}\, dt \;=\; \frac{1}{s}\mathcal{L}(1) \;=\; \frac{1}{s^2}$$

$$(19.6)$$

Thus y(t) = t in the "input domain" becomes Y(s) = $1/s^2$ in the Laplace domain.

You might have noticed that we used the result for the transform of 1 to calculate the transform of t. In general the Laplace transform of t^n has a very simple relationship to the Laplace transform of t^{n-1}.

$$\mathcal{L}(t^n) \;=\; \frac{n}{s}\mathcal{L}(t^{n-1})$$

$$(19.7)$$

This is easy to derive using integration by parts:

$$u = t^n \qquad dv = -\frac{1}{s}e^{-st}(-s)\,dt \qquad and$$

$$du = nt^{n-1}dt \qquad v = -\frac{1}{s}e^{-st}$$

$$L(t^n) = \int_0^\infty t^n e^{-st}dt = -t^n \frac{1}{s}e^{-st}\Big|_0^\infty - \int_0^\infty \left(-\frac{1}{s}\right)e^{-st}nt^{n-1}dt$$

$$(19.8)$$

The exponential e^{st}, for large t, increases faster than any polynomial, including t^n. Thus $\lim\limits_{t\to\infty} -\dfrac{t^n}{e^{st}}\dfrac{1}{s} = 0$. The first term on the right of the equals sign is then zero. Thus:

$$L(t^n) = 0 + \frac{n}{s}\int_0^\infty t^{n-1}e^{-st}\,dt = \frac{n}{s}L(t^{n-1})$$

$$(19.9)$$

Using Eq. 19-6, we can generate the Laplace transforms of all of the powers of t:

$$L(t^2) = \frac{2}{s}L(t) = \frac{2}{s}\frac{1}{s} = \frac{2}{s^2}$$

$$L(t^3) = \frac{3}{s}L(t^2) = \frac{3}{s}\left(\frac{2}{s^2}\right) = \frac{3!}{s^3}$$

$$\vdots$$

$$L(t^n) = \frac{n!}{s^{n+1}}$$

$$(19.10)$$

The Laplace transform of e^{at} is easy to produce in a one-line derivation:

$$L\left(e^{at}\right) = \int_0^\infty e^{at} e^{-st} dt = \int_0^\infty e^{(a-s)t} dt =$$

$$\frac{1}{a-s}\{ e^{(a-s)t} \mid_0^\infty = \frac{1}{s-a}$$

(19.11)

The Laplace transform of sin(at) can be arrived at by integrating by parts twice. The first integration by parts gives us an inner product with cosine:

$$\mathcal{L}(\sin(at)) = \int_0^\infty \sin(at) e^{-st} dt$$

$$u = \sin(at) \qquad dv = e^{-st} dt$$

$$du = a\cos(at) \qquad v = -\frac{1}{s}e^{-st}$$

$$\mathcal{L}(\sin(at)) = -\sin(at)\frac{1}{s}e^{-st} \mid_0^\infty + \frac{a}{s}\int_0^\infty e^{-st} \cos(at) dt$$

$$= \frac{a}{s}\int_0^\infty e^{-st} \cos(at) dt$$

(19.12)

A second integration by parts gives us two terms involving the inner product with sine:

$$u = \cos(at) \qquad dv = e^{-st} dt$$

$$du = -a\sin(at) dt \qquad v = -\frac{1}{s}e^{-st}$$

$$L(\sin(at)) = \frac{a}{s}[-\frac{1}{s}\cos(at)e^{-st} \mid_0^\infty -$$

$$\int_0^\infty (-\frac{1}{s})e^{-st}(-a)\sin(at) dt \ ]$$

(19.13)

$$= \frac{a}{s}[0 - (-\frac{1}{s}) - \frac{a}{s}\int_0^\infty e^{-st} \sin(at) dt \ ]$$

$$= \frac{a}{s^2} - \frac{a^2}{s^2}\int_0^\infty e^{-st} \sin(at) dt$$

Shortcut to Ordinary Differential Equations

The second term on the right of the equals sign involves $\mathcal{L}(\sin(at))$.

$$L(\sin(at)) = \frac{a}{s^2} - \frac{a^2}{s^2}\int_0^\infty e^{-st}\sin(at)\,dt =$$

$$\frac{a}{s^2} - \frac{a^2}{s^2}L(\sin(at))$$

(19.14)

If we combine terms involving $\mathcal{L}(\sin(at))$ we arrive at:

$$\mathcal{L}(\sin(at)) = \frac{a}{s^2} - \frac{a^2}{s^2}\mathcal{L}(\sin(at)) \quad \Rightarrow$$

$$\mathcal{L}(\sin(at))\left\{1 + \frac{a^2}{s^2}\right\} = \frac{a}{s^2}$$

(19.15)

Thus the Laplace transform of $sin(at)$ is:

$$\mathcal{L}(\sin(at)) = \frac{\dfrac{a}{s^2}}{\left\{1 + \dfrac{a^2}{s^2}\right\}} = \frac{a}{s^2 + a^2}$$

(19.16)

When we were performing the above derivation, Eq. 19-21 surreptitiously gave us a relationship between the Laplace transform of cos(at) and the Laplace transform of sin(at).

$$\mathcal{L}(\sin(at)) = \frac{a}{s}\int_0^\infty e^{-st}\cos(at)\,dt = \frac{a}{s}\mathcal{L}(\cos(at))$$

(19.17)

We can use this immediately to obtain $\mathcal{L}(\cos(at))$:

$$\mathcal{L}(\cos(at)) = \frac{s}{a}\mathcal{L}(\sin(at)) = \frac{s}{a}\left(\frac{a}{s^2 + a^2}\right) = \frac{s}{s^2 + a^2}$$

(19.18)

The Laplace transform is a linear transform. This means that the Laplace transform of a sum of functions is the sum of the Laplace transforms of the

130

individual functions. Linearity also means that the Laplace transform of a constant times a function is the constant times the Laplace transform of the function. We can express both aspects of the Linearity of the Laplace transform in a single equation:

$$\mathcal{L}(c_1 y_1(t) + c_2 y_2(t)) = c_1 \mathcal{L}(y_1(t)) + c_2 \mathcal{L}(y_2(t)) \quad (19.19)$$

We can use this property in the development of the Laplace transforms of sinh(at) and cosh(at). In the Laplace domain, the hyperbolic sine and the hyperbolic cosine differ from their non-hyperbolic counterparts by having a minus sign between s^2 and a^2 in the denominator:

$$
\begin{aligned}
\mathsf{L}(\sinh(at)) &= \int_0^\infty \frac{e^{at} - e^{-at}}{2} e^{-st}\, dt \\
&= \frac{1}{2}\int_0^\infty e^{at} e^{-st}\, dt - \frac{1}{2}\int_0^\infty e^{-at} e^{-st}\, dt \\
&= \frac{1}{2}\mathsf{L}(e^{at}) - \frac{1}{2}\mathsf{L}(e^{-at}) \\
&= \frac{1}{2}\left\{ \frac{1}{s-a} - \frac{1}{s+a} \right\} \\
&= \frac{a}{s^2 - a^2}
\end{aligned}
\quad (19.20)
$$

$$
\begin{aligned}
\mathcal{L}(\cosh(at)) &= \int_0^\infty \frac{e^{at} + e^{-at}}{2} e^{-st}\, dt \\
&= \frac{1}{2}\mathcal{L}(e^{at}) + \frac{1}{2}\mathcal{L}(e^{-at}) = \frac{1}{2}\left\{ \frac{1}{s-a} + \frac{1}{s+a} \right\} \\
&= \frac{s}{s^2 - a^2}
\end{aligned}
$$

$$(19.21)$$

We can now tabulate the Laplace transforms we have developed so far and form an initial picture of how some familiar functions appear in the Laplace domain.

Table 19-1: Basic Laplace Transforms.

Input Domain	Laplace Domain
0	0
1	$\dfrac{1}{s}$
t	$\dfrac{1}{s^2}$
t^n	$\dfrac{n!}{s^{n+1}}$
e^{at}	$\dfrac{1}{s - a}$
sin(at)	$\dfrac{a}{s^2 + a^2}$
cos(at)	$\dfrac{s}{s^2 + a^2}$
sinh(at)	$\dfrac{a}{s^2 - a^2}$
cosh(at)	$\dfrac{s}{s^2 - a^2}$

It's interesting that all of the manifold types of functions on the left (polynomials, trigonometric functions, and hyperbolic functions) become ratios of polynomials when transformed to the Laplace domain. These varied functions seem to be on more of a common footing in the Laplace domain, being rational functions in the Laplace domain.

Each row of the table above is an example of a transform pair. The left element is a function in the input domain. The right element is how that function manifests in the

Laplace domain, for example, $\left(\cos(at), \dfrac{s}{s^2 + a^2}\right)$ is a transform pair.

Solving problems involving Laplace transforms entails keeping "two sets of books," one to account for the behavior in the input domain and the other to account for the behavior in the Laplace domain. One can think of Table 19-1 as two accounting books laid side by side. Although we could use Eq. 19-2 and Eq. 19-3 to transform functions back and forth between the input and Laplace domains, we will usually not use Eq. 19-2 and we will almost never use Eq. 19-3. Instead, most of the time we can make the transitions between the two domains by looking up the transform pairs in tables such as Table 19-1.

To illustrate the point, let's use Table 19-1 to transform some other familiar functions into the Laplace domain.

Example 19-1: Determine the Laplace transform of a line.

Solution:

$$y = at + b$$

$$\mathcal{L}(at + b) = a\mathcal{L}(t) + b\mathcal{L}(1) = a\frac{1}{s^2} + b\frac{1}{s}$$

$$Y(s) = a\frac{1}{s^2} + b\frac{1}{s}$$

$$\text{(19.22)}$$

$$\Diamond$$

Here we have transformed a whole equation to the Laplace domain the equation $y = at + b$ becomes $Y(s) = a\dfrac{1}{s^2} + b\dfrac{1}{s}$ when it is transformed to the Laplace domain.

In the next example, the sum of a trigonometric function and a hyperbolic function becomes a simple rational function in the Laplace domain.

Example 19-2: y = cos(at) + sinh(at)

$$L(y) = L(\cos(at)) + L(\sinh(at))$$

Solution:

$$\Downarrow$$

$$Y(s) = \frac{s}{s^2 + a^2} + \frac{a}{s^2 - a^2}$$

$$= \frac{s(s^2 - a^2)}{s^4 - a^4} + \frac{a(s^2 + a^2)}{s^4 - a^4} = \frac{s^3 + a^3}{s^4 - a^4}$$

$$(19.23)$$

$$\Diamond$$

In many instances, Laplace transforms make easy work out of dealing with some rather difficult functions. Functions containing a discontinuity often present challenges. Some discontinuous functions are easy to represent and work with in the Laplace domain, provided they are piecewise continuous. For instance, we can take the Laplace transform of a piecewise defined step. Before we look at any particular step, we need to introduce the unit step function u(t). The function u(t) is equal to zero for t < 0. It is equal to one for t ≥ 0. This allows you to represent a unit step that starts at t = 0 by u(t - a) since t = a makes the argument of *u* zero and thus initiates the start of the upper part of the step. Consider the function:

$$f(t) = \begin{cases} 0, & 0 \le t < a \\ b, & a \le t < \infty \end{cases}$$

$$(19.24)$$

It can be represented as *b u(t - a)* and appears as follows:

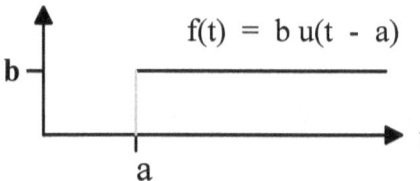

$$f(t) = b\,u(t - a)$$

b

a

t

The Laplace transform of this function is:

$$\mathcal{L}(f(t)) = \int_0^a 0\, dt + \int_a^\infty b e^{-st} dt = \frac{b}{s} e^{-as}$$

$$(19.25)$$

The trick here was to break the original integral, which was from zero to infinity, into an integral from zero to *a* followed by an integral from *a* to infinity. The final result is an amazingly harmless-looking representation of a discontinuous function. What's amazing is that all of the information about the step (start time, the step height, and the shape of the step) is contained in the single term $(b/s)e^{-as}$. This is accomplished without any piecewise definitions. The fact that the discontinuous square wave is transformed to a simple exponential function suggests that problems involving such functions may be much easier to solve in the Laplace domain.

Perhaps the most powerful feature of the Laplace transform is the way it transforms derivatives of functions. Differentiation of y(t) in the input domain manifests as multiplication of Y(s) by s in the Laplace domain with the additional chore of subtracting by y(0).

Table 19-2: Laplace Transform of a Derivative.

Input Domain	Laplace Domain
$\dfrac{dy}{dt}$	$sY(s) - y(0)$

The term being subtracted off is lower-case y(0), which is the <u>input domain</u> initial condition. The object sY(s) - y(0) lives in the Laplace domain notwithstanding the appearance of y(0) which is lower-case. This presents no conceptual problem since y(0) is only a constant, being evaluated at a fixed value of the independent variable.

The next example illustrates using the Laplace transform to solve an initial value problem. Notice how the derivative is reduced to a multiplication by *s* and a subtraction of *y(0)*. The first step is to Laplace-transform the whole differential equation.

Example 19-3: Solve the initial value problem: y' - y = 2cos(5t) with y(0) = 0.

Solution: Applying the Laplace transform to the differential equation, we have:

$$sY(s) - y(0) - Y(s) = 2\left(\frac{s}{s^2 + 5^2}\right)$$

$$(19.26)$$

Plugging in the initial condition y(0) = 0, we have:

$$sY(s) \ - \ Y(s) \ = \ \frac{2s}{s^2 + 5^2}$$

$$Y(s)(s \ - \ 1) \ = \ \frac{2s}{s^2 + 5^2}$$

$$(19.27)$$

This can be solved for Y(s):

$$Y(s) \ = \ \frac{2s}{(s \ - \ 1)(s^2 + 5^2)}$$

$$(19.28)$$

Using partial fractions, we can decompose the right-hand side of this into easier terms:

$$\frac{2s}{(s \ - \ 1)(s^2 + 5^2)} \ = \ \frac{As}{(s^2 + 5^2)} \ + \ \frac{B}{(s^2 + 5^2)} \ + \ \frac{C}{(s \ - \ 1)} \qquad (19.29)$$

Multiplying through by (s - 1)(s² + 5²), we have:

$$2s \ = \ As(s \ - \ 1) \ + \ B(s \ - \ 1) \ + \ C(s^2 + 5^2)$$

$$2s \ = \ (A + C)s^2 \ + \ (B \ - \ A)s \ - \ B \ + \ C5^2$$

$$(19.30)$$

Comparing like terms, this gives three equations in three unknowns:

$$
\begin{array}{rcccccl}
A & & & + & C & = & 0 \\
-A & + & B & & & = & 2 \\
& & -B & + & 25C & = & 0 \\
\end{array}
$$

$$19.31)$$

Whose solution is : $A \ = \ \dfrac{-1}{13}$, $B \ = \ \dfrac{25}{13}$, and $C \ = \ \dfrac{1}{13}$.

Equation 19-28 then becomes:

$$Y(s) \ = \ \left(-\frac{1}{13}\right)\frac{s}{(s^2 + 5^2)} \ + \ \left(\frac{5}{13}\right)\frac{5}{(s^2 + 5^2)} \ + \ \left(\frac{1}{13}\right)\frac{1}{(s \ - \ 1)}$$

$$(19.32)$$

Referring to table 19-1, the terms on the right-hand side can be matched to their time domain equivalents. Thus we can inverse-Laplace transform Eq. 19-42 to obtain the solution:

$$y(t) = -\frac{1}{13}\cos(5t) + \frac{5}{13}\sin(5t) + \frac{1}{13}e^t$$

(19.33)

◊

To derive the Laplace transform of a derivative formula, we use integration by parts to put the derivative onto the exponential:

$$L\left(\frac{df}{dt}\right) = \int_0^\infty \frac{df}{dt}e^{-st}\,dt =$$

$$f(t)e^{-st}\Big|_0^\infty - \int_0^\infty f(t)(-s\,e^{-st})\,dt =$$

(19.34)

$$0 - f(0) - (-s)\int_0^\infty f(t)e^{-st}\,dt = sL(f) - f(0)$$

Again, the primary effect of differentiation is multiplication by s in the Laplace domain. Then the secondary effect is subtraction of the original function evaluated at zero.

The Laplace transform of the second derivative can be derived using the Laplace transform of the first derivative:

$$L\left(\frac{d^2f}{dt^2}\right) = sL\left(\frac{df}{dt}\right) - \frac{df}{dt}(0)$$

$$= s(sL(f) - f(0)) - \frac{df}{dt}(0)$$

(19.35)

$$= s^2L(f) - s\,f(0) - \frac{df}{dt}(0)$$

Now we have two derivatives that we can tabulate.

Shortcut to Ordinary Differential Equations

Table 19-3: Laplace Transforms of Derivatives.

Input Domain	Laplace Domain
$\dfrac{dy}{dt}$	$sY(s) - y(0)$
$\dfrac{d^2y}{dt^2}$	$s^2Y(s) - s\,y(0) - y'(0)$

Solving the harmonic oscillator problem was quite a chore using the *in situ* methods of the previous articles. Using the Laplace transform, it's a breeze.

Example 19-4: Harmonic oscillator. Solve $\ddot{y} + \omega^2 y = 0$ with $y(0) = 1$ and $y'(0) = 0$.
(ω is a constant parameter, "frequency.")

Solution: Apply the Laplace transform to the entire equation:

$$\mathcal{L}\left(\frac{d^2y}{dt^2} + \omega^2 y \right) = \mathcal{L}(0)$$

(19.36)

The next step is to use the linearity of the Laplace transform to distribute $\mathcal{L}$.

$$\mathcal{L}\left(\frac{d^2y}{dt^2} \right) + \omega^2 \mathcal{L}(y) = \mathcal{L}(0)$$

(19.37)

Making use of our tables, we can write the Laplace domain equation:

$$s^2Y(s) - s\,y(0) - y'(0) + \omega^2 Y(s) = 0$$

(19.38)

Using the initial conditions $y(0) = 1$ and $y'(0) = 0$, we have:

$$s^2Y(s) - s + \omega^2 Y(s) = 0$$

(19.39)

We can solve this for Y(s):

$$Y(s) = \frac{s}{s^2 + \omega^2}$$

$$(19.40)$$

We have solved the problem in the Laplace domain by isolating the independent variable Y and expressing in in terms of the dependent variable s. In other words, we have found Y = Y(s). Now we need to transform our answer back to the "real world" domain where

it appears in the form y = y(t). If we look at Table 19-1, we see that $\dfrac{s}{s^2 + a^2}$

corresponds to cos(at) in the input domain Thus the inverse-transform of

$$Y(s) = \frac{s}{s^2 + \omega^2}$$ is cos(ωt).

$$y(t) = \cos(\omega t)$$

$$(19.41)$$
◊

Being able to effect differentiation via multiplication goes both ways. We can take a derivative with respect to s in the Laplace domain by multiplying by $-t$ in the input domain:

$$\mathcal{L}(-t\, f(t)) = \frac{d}{ds} F(s)$$

$$(19.42)$$

Leibniz's rule allows us to take the derivative into the integral sign when we differentiate the following with respect to s:

$$F(s) = \int_0^\infty f(t) e^{-st} dt$$

$$(19.43)$$

Thus we have:

$$\frac{d}{ds} F(s) = \int_0^\infty \frac{\partial}{\partial s} f(t) e^{-st} dt = \int_0^\infty f(t)(-t) e^{-st} dt = \mathsf{L}(-t\, f(t))$$

$$(19.44)$$

We can use this to determine the Laplace transforms of some harder functions.

Shortcut to Ordinary Differential Equations

Example 19-5: Determine the Laplace transform of $-t\cos(at)$.

The Laplace transform of cos(at) is:

$$\frac{s}{s^2 + a^2} \tag{19.45}$$

Multiplication by $-t$ causes differentiation in the Laplace domain:

$$\frac{1}{s^2 + a^2} + s(-1)\frac{2s}{\left(s^2 + a^2\right)^2} \tag{19.46}$$

Which becomes:

$$\frac{a^2 - s^2}{\left(s^2 + a^2\right)^2} \tag{19.47}$$

◊

EXERCISES

Exercise 19-1: Determine Y(s), the Laplace transform of the following time domain function y(t):

$$y(t) = 2\sin(t)\cos(t) \tag{19.48}$$

Exercise 19-2: Determine Y(s), the Laplace transform of y(t), where y(t) is as follows:

$$y(t) = t^6\sin^2(7t) + t^6\cos^2(7t) \tag{19.49}$$

Exercise 19-3: Determine y(t), the <u>inverse</u> Laplace transform of Y(s):

$$Y(s) = \frac{1}{s} - \frac{1}{s^2 + 1} \tag{19.50}$$

Exercise 19-4: Use Laplace transforms to solve the following initial value problem.

140

$$2\frac{dy}{dt} + y = 0 \qquad y(0) = -3.$$
(19.51)

Exercise 19-5: Solve the initial value problem:

$$y'' + 2y' + 2y = 2 \qquad \text{with} \qquad y(0) = 1 \text{ and } y'(0) = 0$$
(19.52)

Exercise 19-6: Solve $\ddot{y} + \omega^2 y = 0$ with $y(0) = 0$ and $y'(0) = 1$.

Exercise 19-7: A step function is defined as in Equation 19-24:

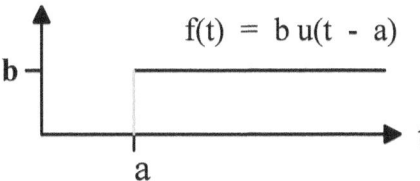

Determine the Laplace transform of the *derivative* of this function. Let $b = 1$.

Exercise 19-8: Determine the Laplace transform of $t^2 sin(at)$.

Article 20. Laplace Transform Translation Theorems

Pierre-Simon Laplace
(1749-1827)

Pierre-Simon Laplace had a profound influence on mathematics, astronomy, and physics and wrote huge texts on these subjects. One of the great tools that was developed by Laplace is the Laplace transform, which was first published in 1782. This transform was used extensively in a huge treatise in probability that Laplace wrote. Laplace's great works featured not only Laplace's contributions to math and science, but also results developed by other geniuses. Unfortunately, Laplace sometimes presented other mathematicians' material in a way that suggested the offerings were all his own. Notwithstanding the problem of omitted acknowledgments, the contributions of Laplace to science were enough for him to be remembered as "The Newton of France."

In the previous article, we became acquainted with the Laplace transforms of some basic functions. In this article we will learn how to extend those results so that we are able to transform many more functions. Our mainstay approach will be to apply translations (shifts) to the functions we are dealing with in either the input domain or the Laplace domain.

When working with Laplace transform problems, we think of functions f(t) as belonging to an *input domain* and the Laplace transformed version F(s) as belonging to the Laplace domain. We have already seen that differentiation in the input domain manifests as multiplication by the independent variable in the Laplace domain (less the value of the original function at zero). We've seen also that multiplication by -t in the input domain manifests as differentiation with respect to s in the Laplace domain. Multiplication of a function y(t) by $e^{\xi t}$ in the input domain has an interesting way of showing up in the Laplace transform domain: It causes the function in the Laplace domain to shift to the right along the s axis by the amount ξ. In other words, if the Laplace transform of f(t) is F(s), then the Laplace transform of $e^{\xi t}$ f(t) is F(s - ξ).

$$\mathcal{L}\left(e^{\xi t} f(t)\right) = \int_0^\infty e^{\xi t} f(t) e^{-st} dt$$

$$= \int_0^\infty f(t) e^{-(s-\xi)t} dt$$

$$= F(s - \xi) \tag{20.1}$$

The insight here was to recognize, in the second integral, that $s - \xi$ plays the role of a new Laplace parameter, i.e., $s_2 = s_1 - \xi$. The Laplace domain function F(s) is shifted to the right by the amount ξ in response to the input domain event of multiplication by $e^{\xi t}$. This is also referred to as a translation by the amount ξ.

Example 20-1: Determine the Laplace transform of $e^{3t}\sin(t)$.

Solution: Let sin(t) be f(t), a function in the input domain. Now consider the effect of multiplication by e^{3t}. It's important to keep straight that if ξ is positive, then s is replaced by $s - \xi$ in the Laplace domain. We don't really need to *calculate* the Laplace transform all over from scratch, we simply need to replace s with s - 3 to account for the fact that sin(t) is being multiplied by e^{3t}. We have:

$$\sin(t) \qquad \Leftrightarrow \qquad \frac{1}{s^2 - 1}$$

$$e^{3t}\sin(t) \qquad \Leftrightarrow \qquad \frac{1}{(s-3)^2 - 1} \tag{20.2}$$

Thus:

$$\mathcal{L}\left\{e^{3t}\sin(t)\right\} \qquad \Leftrightarrow \qquad \frac{1}{(s-3)^2 - 1} \tag{20.3}$$

$$\Diamond$$

The style of solution is a little different here than what we may be used to. In Eq. 20-2, we are basically saying, "If sin(t) is associated with $1/(s^2 - 1)$ then $e^{3t}\sin(t)$ is associated with $1/((s-3)^2 - 1)$." By "associated" we mean "is a transform pair with."

Translation theorems help us quickly obtain Laplace transforms without having

144

to go back to the original definition of the Laplace transform and crank the integral. Recall that a translation of a function $\Psi(s)$ can be realized by replacing s with s - a. The effect is to shift $\Psi(s)$ to the right by the amount a. $\Psi(s - a)$ is the resulting shifted version.

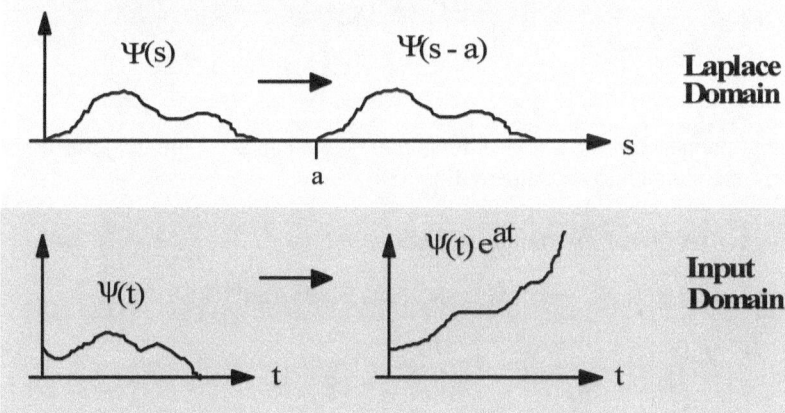

But shift by the amount a in the *Laplace* domain effects multiplication by e^{at} in the input domain. Think of the two operations occurring in the two domains as happening in parallel. The first translation theorem says: If we shift by a in Laplace domain, we multiply by e^{at} in the input domain and if we multiply by e^{at} in the input domain, we shift by a in the Laplace domain.

> *First Translation Theorem :*
>
> If $\mathcal{L}\{f(t)\} = F(s)$ and a is any real number,
>
> $\mathcal{L}\{e^{at} f(t)\} = F(s - a)$

$$(20.4)$$

•

Let's look at a couple of examples:

Example 20-2: Determine the Laplace transform of $t^{10}e^{-6t}$.

$$\mathcal{L}\{t\} = \frac{1}{s^2} \Rightarrow \mathcal{L}\{t^{10} e^{-6t}\} = \frac{10!}{(s + 6)^{11}}$$

$$(20.5)$$

◊

Example 20-3: Determine the Laplace transform of $e^{2t}(t - 1)^2$:

145

$$L\{e^{2t}(t-1)^2\} = L\{t^2 e^{2t} - 2t e^{2t} + e^{2t}\}$$

$$= L\{t^2 e^{2t}\} - 2L\{t e^{2t}\} + L\{1 e^{2t}\} \qquad (20.6)$$

$$= \frac{2}{(s-2)^3} - \frac{2}{(s-2)^2} + \frac{1}{s-2}$$

◊

The input domain and the Laplace domain are "two hands clapping." Thus there is an inverse theorem to accompany Eq. 20-4.

Inverse First Translation Theorem :

If $L^{-1}\{F(s)\} = f(t)$ and a is any real number,

$$L^{-1}F(s-a) = e^{at}f(t)$$

$$(20.7)$$

Let's use this theorem to see how we can go from the Laplace domain back to the input domain.

Example 20-4: Determine the inverse Laplace transform of $1/(s^2 + 2s + 5)$.

Solution: The trick here is to complete the square in the denominator.

$$L^{-1}\left\{\frac{1}{s^2 + 2s + 5}\right\} = L^{-1}\left\{\frac{1}{s^2 + 2s + 1 + 4}\right\}$$

$$= L^{-1}\left\{\frac{1}{(s+1)^2 + 4}\right\}$$

$$(20.8)$$

Next, we can factor out ½ to put this in the sine form:

$$\mathcal{L}^{-1}\left\{\frac{1}{s^2 + 2s + 5}\right\} = \mathcal{L}^{-1}\left\{\frac{1}{(s + 1)^2 + 4}\right\}$$

$$= \frac{1}{2}\mathcal{L}^{-1}\left\{\frac{2}{(s + 1)^2 + 2^2}\right\}$$

$$= \frac{1}{2}e^{-t}\sin(2t)$$

(20.9)

◊

Example 20-5: Solve $y' + y = 3e^{2t}$ with $y(0) = 0$.

Solution: Laplace transforming, we obtain:

$$s\,Y(s) - y(0) + Y(s) = \frac{3}{s - 2} \qquad \text{Or :}$$

$$Y(s)(s + 1) = \frac{3}{s - 2}$$

(20.10)

Thus:

$$Y(s) = \frac{3}{(s + 1)(s - 2)}$$

(20.11)

Using partial fractions, this becomes:

$$Y(s) = \frac{-1}{(s + 1)} + \frac{1}{(s - 2)}$$

(20.12)

Now, taking the inverse Laplace transform, we have our answer:

$$y(t) = -e^{-t} + e^{2t}$$

(20.13)

◊

In the next example, we apply Laplace transforms to the solution of an LRC circuit with an applied voltage.

Example 20-6: LRC Circuit with zero applied voltage. The simplest possible LRC circuit consists of an inductor L, a capacitor C, and a resistor R. The current through the circuit is j = j(t). The initial conditions are j(0) = 0 and j'(0) = 1. Solve for the current as a function of time.

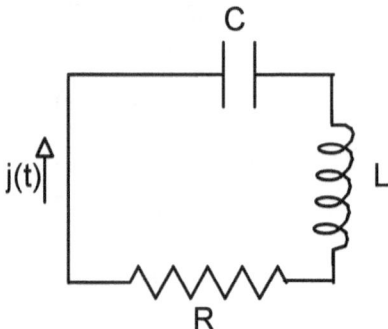

Determining an equation to describe the circuit requires adding up the voltages across each of these elements. The current j(t) is the rate at which charge q is flowing per unit time, that is, j(t) = dq/dt. The voltage drop across the inductor is V_L = L(dj/dt). An inductor has a magnetic field in it which "pushes back" against any change in current, dj/dt. The larger the inductance, the larger the voltage drop for a given change in current. A resistor is a device that "resists" the flow of electrons. The voltage drop across the resistor is V_R = Rj(t). This reflects the fact that, the larger the resistance R, the larger the voltage needed to maintain a certain current flow through the resistor. A capacitor is a temporary charge storage device. Equal and opposite charges of amount q occur on plates separated from each other by a gap. The voltage drop across a capacitor is V_C = q/C. For a small capacitance, a large voltage is required to keep the charges separate. For a large charge, a large voltage is present. Kirchoff's Law says that these voltage drops must add up to the applied voltage which, in this case, is 0.

$$V_L + V_R + V_C = V_{\text{APPLIED}} \qquad (20.14)$$

Plugging in, we have:

$$L\frac{dj}{dt} + Rj + \frac{1}{C}q = 0 \qquad (20.15)$$

The current j is the time derivative of charge: j(t) = dq/dt. Differentiating Eq. 20-15, we obtain a second-order ordinary differential equation:

$$L\frac{d^2 j}{dt^2} + R\frac{dj}{dt} + \frac{1}{C}j = 0$$
$$\text{(20.16)}$$

Remembering that R, L, and C are constants, we are now ready to apply the Laplace transform to the whole equation. Be careful not to confuse the inductance L with the Laplace transform $\mathcal{L}$:

$$L\{s^2 J - s\, j(0) - j'(0)\} + R\{s\, J - j(0)\} + \frac{1}{C}J = 0 \quad \text{(20.17)}$$

Plugging in our initial conditions, this becomes:

$$L s^2 J - L + R s J + \frac{1}{C}J = 0$$
$$\text{(20.18)}$$

Separating variables, we have:

$$J = \frac{L}{L s^2 + R s + \frac{1}{C}} = \frac{1}{s^2 + \frac{R}{L}s + \frac{1}{LC}}$$
$$\text{(20.19)}$$

We can complete the square in the denominator:

$$J = \frac{1}{s^2 + \frac{R}{L}s + \frac{R^2}{4L^2} + \frac{1}{LC} - \frac{R^2}{4L^2}}$$

$$= \frac{1}{\left(s + \frac{R}{2L}\right)^2 + \frac{1}{LC} - \frac{R^2}{4L^2}}$$
$$\text{(20.20)}$$

If we can put this in a form where its recognizable as a translated sine function:

$$J = \frac{1}{\sqrt{\frac{1}{LC} - \frac{R^2}{4L^2}}} \cdot \frac{\sqrt{\frac{1}{LC} - \frac{R^2}{4L^2}}}{\left\{\left(s + \frac{R}{2L}\right)^2 + \frac{1}{LC} - \frac{R^2}{4L^2}\right\}}$$
$$\text{(20.21)}$$

Thus:

$$j(t) = \frac{1}{\sqrt{\dfrac{1}{LC} - \dfrac{R^2}{4L^2}}} e^{-\frac{R}{2L}t} \sin(\sqrt{\frac{1}{LC} - \frac{R^2}{4L^2}}\, t)$$

(20.22)

It may seem like a bit of a leap from Eq. 20-21 to Eq. 20-22; however, if we look at the components, it's straightforward. The right-most term is in the form

$$\frac{a}{\left(s - \frac{R}{2L}\right)^2 + a^2}$$ and the translation R/2L in the Laplace domain was accounted

for in the input domain as multiplication by $e^{-(R/2L)t}$. Thus we have sin(at) with $e^{-(R/2L)t}$ and a constant term in front.

Eq. 20-22 describes a current that oscillates but decays over time so long as the resistance R is greater than zero and less than $2(L/C)^{\frac{1}{2}}$. If R = 0, the solution reduces to the harmonic oscillator solution for which the oscillations do not decay:

If R = 0.

$$j(t) = \sqrt{LC} \sin(\frac{1}{\sqrt{LC}}\, t)$$

(20.13)

◊

If you listen to the din of system engineers discussing digital signal processing, you are likely to hear the word "impulse response" being uttered. What engineers are referring to when they speak about impulse response is quite literally the *response* of a system when an *impulse* function is applied to the system. For a broad class of systems, how the system responds when a narrow impulse function is applied, provides a comprehensive characterization of all of the system's attributes. In particular, for linear systems, if the response of the system to an impulse is known, then the response of that system to any input can be predicted.

When dealing with functions of a continuous variable, an object known as a Dirac delta function is used to represent an impulse. The attributes of a Dirac delta function are these: 1.) It has constant area; 2.) It has infinite height; 3.) It has zero width; and 4.) It is centered over the value that makes its argument zero.

150

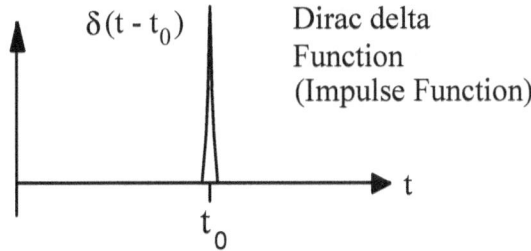

We can construct a Dirac delta function from the step function we worked with in Article 19. Recall that we discussed the following step function:

Consider the function:

$$f(t) = \begin{cases} 0, & 0 \le t < a \\ b, & a \le t < \infty \end{cases}$$

(20.14)

It can be represented as $b\,u(t - a)$ and appears as follows:

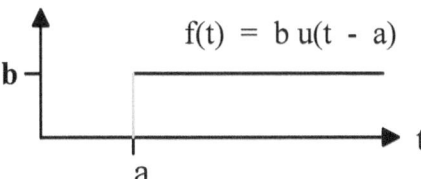

As we have already seen, the Laplace transform of this function is:

$$\mathcal{L}(f(t)) = \int_0^a 0\,dt + \int_a^\infty b e^{-st}\,dt = \frac{b}{s} e^{-as}$$

$$F(s) = \frac{b}{s} e^{-as}$$

(20.15)

If we let $a = t_0$ and $b = 1/h$ then the step, which we will call $y_1(t)$ has the following appearance:

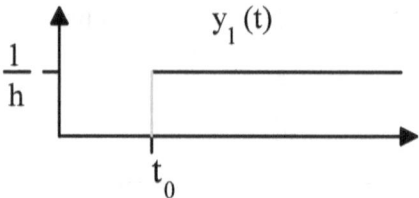

Plugging into Eq. 20-16, the Laplace transform of $y_1(t)$ is:

$$Y_1(s) = \frac{1}{sh}e^{-t_0 s}$$

(20.16)

We can make another step function, $y_2(t)$, shifted over by the amount h and having the same height 1/h:

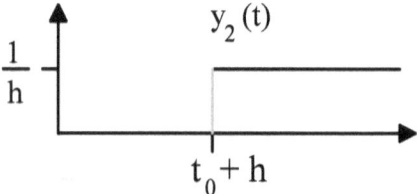

Again plugging into Eq. 20-16, the Laplace transform of this is:

$$Y_2(s) = \frac{1}{sh}e^{-(t_0 + h)s}$$

(20.17)

We can form an impulse of unit area, width h, and height 1/h by forming y(t) = $y_1(t)$ - $y_2(t)$. Behold:

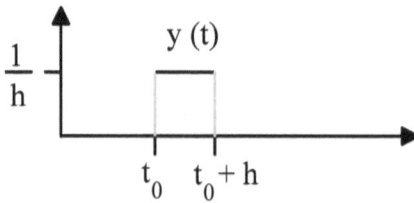

Since the Laplace transform is a linear transform, we can determine the Laplace transform of y(t) easily since we know the Laplace transforms of $y_1(t)$ and $y_2(t)$.

$$Y(s) = L\{y_1(t) - y_2(t)\} = L\{y_1(t)\} - L\{y_2(t)\}$$

$$= Y_1(s) - Y_2(s) = \frac{1}{sh}e^{-t_0 s} - \frac{1}{sh}e^{-(t_0 + h)s} \qquad (20.18)$$

$$= \frac{1}{sh}e^{-t_0 s}\left(1 - e^{-hs}\right) = e^{-t_0 s}\left\{\frac{1 - e^{-hs}}{hs}\right\}$$

The Dirac delta function is the limiting case of this function y(t) for infinite height and zero width. This occurs as h approaches zero:

$$\text{Dirac delta function}:$$

$$\delta(t - t_0) = \lim_{h \to 0} y(t) \qquad (20.19)$$

Again, the spike of the Dirac delta function occurs at $t = t_0$ and this is the value of t that makes the argument of the dirac delta function zero. The Laplace transform of the Dirac delta function can be approached by taking the Laplace transform of Eq. 20-18 and then letting h again approach zero.

$$L\{\delta(t - t_0)\} = \lim_{h \to 0} e^{-t_0 s}\left\{\frac{1 - e^{-hs}}{hs}\right\}$$

$$= e^{-t_0 s}\lim_{h \to 0}\left\{\frac{1 - e^{-hs}}{hs}\right\} \qquad (20.20)$$

Using l'Hospital's rule, being careful to differentiate with respect to s, this limit can be evaluated:

$$L\{\delta(t - t_0)\} = e^{-t_0 s}\lim_{h \to 0}\left\{\frac{h\,e^{-hs}}{h}\right\} = 1\,e^{-t_0 s} \qquad (20.21)$$

We are now in a position to extend our concept of a derivative. In particular, we are now able to take the derivative of a step.

Example 20-7: Derivative of a step. Take the derivative of the following step function:

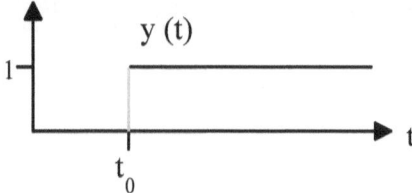

Solution:
The step appears in the Laplace domain as:

$$\mathcal{L}\left\{\frac{dy}{dt}\right\} = s\,\mathcal{L}(y) - y(0) = s\frac{1}{s}e^{-t_0 s} - 0 = e^{-t_0 s}$$

$$(20.22)$$

But this is the Laplace transform of the Dirac delta function centered at t_0. Thus:

$$\mathcal{L}\left\{\frac{dy}{dt}\right\} = \delta(t - t_0)$$

$$(20.23)$$

◊

There is a second translation theorem for Laplace transforms. If a function F(s) in the Laplace domain is multiplied by e^{-as}, the effect in the input domain is twofold: the function f(t) is shifted to the right by the amount a and then this shifted function is set equal to zero for any times less than a. That is, f(t) is replaced by f(t - a)u(t - a):

Second Translation Theorem :

If $F(s) = \mathcal{L}\{f(t)\}$ and $a > 0,$ then

$$\mathcal{L}\{f(t - a)u(t - a)\} = e^{-as}F(s)$$

$$(20.24)$$

Thus multiplication by e^{-as} has the following appearance in the time domain:

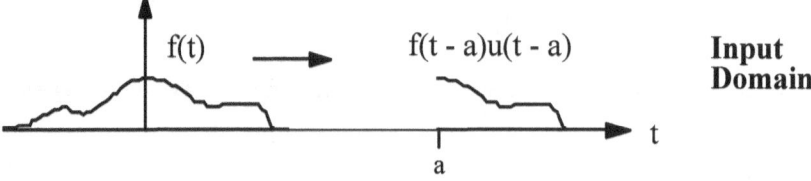

Example 20-8: Struck Harmonic Oscillator. Solve $y'' + 16y = \delta(t - 2\pi)$ with initial conditions $y(0) = 0$ and $y'(0) = 0$.

Solution: This is a system that is at rest until it is struck by an impulse at time $t = 2\pi$. If we transform the equation, we have:

$$s^2 Y(s) - s\,y(0) - y'(0) + 16 Y(s) = e^{-2\pi s} \qquad (20.25)$$

The initial conditions are both zero; therefore, this reduces to:

$$\left(16 + s^2\right)Y(s) = e^{-2\pi s} \qquad (20.26)$$

Or:

$$Y(s) = \frac{1}{4}\frac{4e^{-2\pi s}}{\left(s^2 + 16\right)} \qquad (20.27)$$

Now this is in the form that sine takes when transformed to the Laplace domain. In this case, it is shifted in the input domain by 2π (and truncated for time $t < 2\pi$) and has frequency $a = 4$. So, transforming back to the input domain, the solution is:

$$y = \frac{1}{4}\sin(4t)u(t - 2\pi) \qquad (20.28)$$

Or:

$$y = \begin{cases} 0, & t < 2\pi \\ \frac{1}{4}\sin(4t), & t \geq 2\pi \end{cases} \qquad (20.29)$$

We can illustrate the solution as follows:

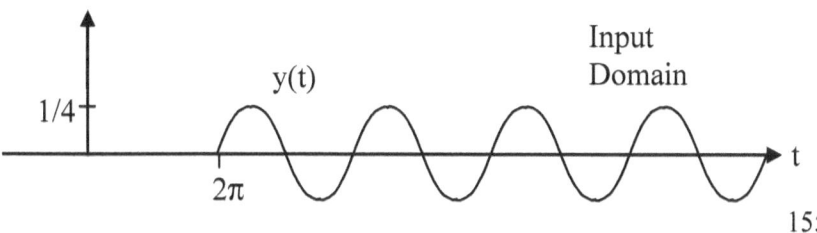

The system is quiet until it is hit with an impulse at time $t = 2\pi$ at which time it oscillates with amplitude ¼.

◊

_____

EXERCISES

Exercise 20-1: Use Laplace transforms to solve the initial value problem $y'' + 4y' + 13y = 0$ with initial conditions $y(0) = 1$ and $y'(0) = -2$.

Exercise 20-2: Use Laplace transforms to solve the initial value problem $y'' + y = e^t$ with initial conditions $y(0) = 1$ and $y'(0) = 1$.

Exercise 20-3: Solve, using Laplace transforms, the initial value problem: $y'' + 5y' + 6y = 0$ with initial conditions $y(0) = 0$ and $y'(0) = 1$.

Exercise 20-4: Solve the initial value problem: $\ddot{y} + 9y = \delta(t - 3\pi/2)$ with intial conditions $y(0) = 0$ and $y'(0) = 0$.

Exercise 20-5: An LRC circuit has an applied voltage $e^{-\omega t}$. We wish to solve for the current $j = j(t)$. The initial conditions are $j(0) = 0$ and $j'(0) = 0$.

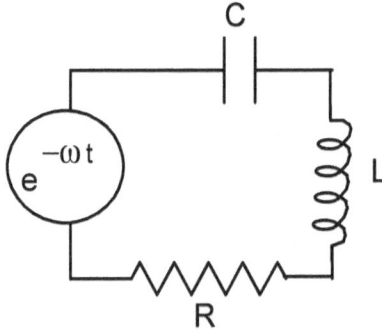

Applying Kirchoff's Law, the voltages add up as follows:

$$L\frac{dj}{dt} + Rj + \frac{1}{C}Q = e^{-\omega t}$$

(20.30)

The current j is the time derivative of charge: $j(t) = dQ/dt$. Differentiating Eq.

156

20-30, we obtain a second-order ordinary differential equation:

$$L\frac{d^2 j}{dt^2} + R\frac{dj}{dt} + \frac{1}{C}j = -\omega e^{-\omega t}$$

(20.31)

Show that the Laplace transform of the current, J(s) is:

$$J(s) = -\frac{\omega}{(\omega + s)}\frac{1}{(Ls^2 + Rs + \frac{1}{C})}$$

(20.32)

Exercise 20-6: For the previous problem, solve for j = j(t) if ω = 1, L = 1, R = 2, C = ½.

Appendix 1: Answers to Exercises

Exercise 2–1:

$$y + \frac{t_{\text{half}}}{\ln 2}\dot{y} = y_0 e^{-t\frac{\ln 2}{t_{\text{half}}}} + \frac{t_{\text{half}}}{\ln 2}\left(-y_0\frac{\ln 2}{t_{\text{half}}}\right)e^{-t\frac{\ln 2}{t_{\text{half}}}} = 0$$

Exercise 2–2: $\quad y'' + y =$
$$-C_1\cos x - C_2\sin x + C_1\cos x + C_2\sin x = 0$$

Exercise 2–3:

$$v + \frac{c^2}{y_0^2}y_0\sqrt{1 - \frac{v^2}{c^2}}\,\frac{y_0}{2}\left(1 - \frac{v^2}{c^2}\right)^{-\frac{1}{2}}\left(-\frac{2v}{c^2}\right) = 0$$

Exercise 3–1: $\quad y = Ce^{-\frac{t}{b}}$

Exercise 3–2: $\quad y = 2\left(\frac{1 + Ce^{4x}}{1 - Ce^{4x}}\right)$

Exercise 3–3: $\quad y = y_0\sqrt{1 - \frac{v^2}{c^2}}$

Exercise 3–4: $\quad y = \dfrac{C}{x^2 - \pi}$

Exercise 4–1:

$$\frac{1}{6}e^x - 5Ce^{-5x} + 5\left\{\frac{1}{6}e^x + Ce^{-5x}\right\} = e^x$$

Exercise 4 − 2:　　$\dfrac{\partial y}{\partial t} = -e^{-t}\cos x$　　$\dfrac{\partial y}{\partial x} = -e^{-t}\sin x$

$$\dfrac{\partial^2 y}{\partial x^2} = -e^{-t}\cos x = \dfrac{\partial y}{\partial t}$$

Exercise 4 − 3:　　1st - order; 4th - order; 7th - order

Exercise 4 − 4:　　linear; non - linear; linear; linear

Exercise 5 − 1:　　$y = \dfrac{1}{t^2 - 3} - 4$

Exercise 5 − 2:　　$y = 6m$

Exercise 6 − 1:　　No guarantee of existence or uniqueness.

Exercise 6 − 2:

Both $f(x, y)$ and $\dfrac{\partial f}{\partial y}$ are continuous in the neighborhood of (x_0, y_0).

Exercise 7 − 1:　　$y = -\dfrac{x^2}{2} + C$

Exercise 7 − 2:　　$C = x^3 y - \dfrac{1}{3} y^3$

Exercise 7 − 3:　　$y^2 + t \sin y = C$

Exercise 7 − 4:　　$y = \cot x$

Exercise 7–5:

Not exact : Solve using separable variables. $y = e^{\frac{C}{x}}$

Exercise 8–1: $y = \dfrac{1}{4}e^{2x} + Ce^{-2x}$

Exercise 8–2: $y = xe^{-\tan x} + Ce^{-\tan x}$

Exercise 8–3: $y = \dfrac{1}{2}\{\sin x - \cos x\} + Ce^{-x}$

Exercise 8–4: $y = e^{-x^2}(x + C)$

Exercise 8–5: $y = \dfrac{E}{R}\left(1 - e^{-\frac{R}{L}t}\right)$

Exercise 9–1: $y = \pm\sqrt{\dfrac{C}{x} - \dfrac{x^2}{3}}$

Exercise 9–2: $x\left(1 + \sin\left(\dfrac{y}{x}\right)\right) = C\cos\left(\dfrac{y}{x}\right)$

Exercise 9–3: $\dfrac{x}{\sin(t/x)} = C$

Exercise 9–4: $y = Cx^{-3} - \dfrac{x}{4}$

Exercise 10–1: $y = x^{-1}(6x + Cx^3)^{-1/3}$

$$\text{Exercise } 10-2: \quad y = \frac{\sqrt{\sin(x^4)+C}}{x}$$

$$\text{Exercise } 10-3: \quad y = x^{-1}\left(2\sin(1/x)+C\right)^{-1/2}$$

$$\text{Exercise } 11-1: \quad \|z\|^2 = 5$$

$$\text{Exercise } 11-2: \quad \cos\left(\frac{\ln i}{i}\right) = 0$$

$$\text{Exercise } 11-3: \quad 1^{\frac{1}{4}} = \{i,-1,-i,1\}$$

$$\text{Exercise } 11-4: \quad h(t)^* h(t) = 1$$

$$\text{Exercise } 11-5: \quad \|h(t)\| = \sqrt{2\pi}$$

$$\text{Exercise } 11-6: \quad FT(e^{j\omega t}) = 2\pi$$

$$\text{Exercise } 11-7: \quad i^i = e^{-(\pi/2 + 2\pi)} = .000388$$
$$i^i = e^{-(\pi/2 + 4\pi)} = 7.25 \times 10^{-7}$$

Exercise $12-1$:

$$\sqrt{2}\cos\left(\omega t - \frac{\pi}{4}\right) = \cos(\omega t) + \sin(\omega t)$$

$$\text{Exercise } 12-2: \quad W = 0$$

$$\text{Exercise } 12-3: \quad \frac{d^2 y_h}{dt^2} + \frac{1}{LC} y_h = 0$$

Exercise 12 – 4 :

$$\left(-\omega_0^2 + \frac{1}{LC}\right)\frac{\cos(\omega_0 t)}{\left(1 - \omega_0^2 LC\right)}\omega_0 C = \frac{\omega_0}{L}\cos(\omega_0 t)$$

Exercise 12 – 5 : $j_g = j_h + j_p =$

$$C_1 \cos\left(\sqrt{LC}t\right) + C_2 \sin\left(\sqrt{LC}t\right) + \frac{\omega_0 C}{1 - \omega_0^2 LC}\cos(\omega_0 t)$$

Exercise 12 – 6 : $C_{resonance} = \dfrac{1}{\omega_0^2 L}$

Exercise 12 – 7 : $\omega_1 = \dfrac{2\pi}{L}$

Exercise 12 – 8 : $\omega_1 = \dfrac{4\pi}{L}$

Exercise 13 – 1 : $y_g = C_1 e^x + C_2 e^{2x}$

Exercise 13 – 2 : $y_g = e^t\left(C_1 \cos(2t) + C_2 \sin(2t)\right)$

Exercise 13 – 3 : $y_g = C_1 e^{4x} + C_2 x e^{4x}$

Exercise 13 – 4 : $y = e^{\pi x}$

Exercise 14 – 1 : $y = C_0 \dfrac{e^{C_0 x + C_1} + 1}{e^{C_0 x + C_1} - 1}$

Exercise 14 – 2: $\quad y_g = C_1 e^{\frac{1}{2}x} + C_2 e^{-\frac{1}{2}x}$

Exercise 14 – 3: $\quad y_p = \dfrac{x^3}{12}$

Exercise 15 – 1: $\quad y_g = C_1 \cos 2x + C_2 \sin 2x + \sin x$

Exercise 15 – 2:

$$y_g = C_1 e^{-\frac{1}{2}x} + C_2 e^{-x} - \frac{1}{10}\cos x + \frac{3}{10}\sin x$$

Exercise 15 – 3: $\quad y_g = C_1 e^{3x} + C_2 e^{-x} + x^2 e^{3x}$

Exercise 16 – 1: $\quad y_p = -\dfrac{1}{9}\cos(3x)\ln\left|\sec(3x) + \tan(3x)\right|$

Exercise 16 – 2: $\quad y_g = C_1 e^x + C_2 e^{-x} + \dfrac{1}{3}e^{2x}$

Exercise 16 – 3: $\quad y_g = C_1 e^{-x} + C_2 x e^{-x} + \dfrac{1}{16}e^{3x}$

Exercise 16 – 4:

$$y_g = C_1 \cos(x) + C_2 \sin(x) - x\cos(x) + \sin(x)\ln\left|\sin x\right|$$

Exercise 17 – 1: $\quad y_g = C_1 x^{-2} + C_2 x^{-10}$

Exercise 17 – 2: $\quad y_g = C_1 x^{-1} + C_2 x^{-1}\ln x + \pi$

Exercise 17 – 3 :

$$y_g = e^{-\frac{1}{2}\ln x}\left[C_1 \cos(\ln x) + C_2 \sin(\ln x)\right]$$

Exercise 18 – 1 : $y_g = C_1 \cos(7x) + C_2 \sin(7x)$

Exercise 18 – 2 : $y = 1$

Exercise 18 – 3 : $y = x$

Exercise 18 – 4 : $y = 1 - 2x^2$

Exercise 19 – 1 : $Y(s) = \dfrac{2}{s^2 + 4}$

Exercise 19 – 2 : $Y(s) = \dfrac{720}{s^7}$

Exercise 19 – 3 : $y(t) = 1 - \sin(t)$

Exercise 19 – 4 : $y(t) = -3e^{-\frac{t}{2}}$

Exercise 19 – 5 : $y(t) = 1$

Exercise 19 – 6 : $y(t) = \dfrac{1}{\omega}\sin(\omega t)$

Exercise 19 – 7 : $\dfrac{df}{dt} = e^{-at}$

Exercise 19 – 8 : $Y(s) = -2a\dfrac{\left(a^2 - 3s^2\right)}{\left(a^2 + s^2\right)^3}$

Exercise 20 – 1 : $y(t) = e^{-2t}\cos 3t$

Exercise 20 – 2 : $y(t) = \dfrac{1}{2}\left(\cos t + \sin t + e^t\right)$

Exercise 20 – 3 : $y(t) = e^{-2t} - e^{-3t}$

Exercise 20 – 4 : $y(t) = \sin(3t)\,u(t - \dfrac{3\pi}{2})$

Exercise 20 – 5 :

$$J(s) = -\dfrac{\omega}{\left(\omega + s\right)}\dfrac{1}{(Ls^2 + Rs + \dfrac{1}{C})}$$

Exercise 20 – 6 : $j(t) = e^{-t}\left(1 - \cos t\right)$

Appendix 2: Timeline of Ordinary Differential Equations

Most of the techniques we use in this course have names that attach to them. The following timeline indicates when these mathematicians lived and where they were from. Some of these people were rather worldly so it's difficult to pin them down to a single country. Euler, for example, although Swiss, decamped to Russia and worked there for many years.

	1600	1700	1800	1900	2000
England	Newton	Taylor		Cayley	Dirac
Germany	Leibniz		Bessel	Frobenius / Gauss / Grassmann	Hilbert
Switzerland	James Bernoulli	John Bernoulli / Daniel Bernoulli / Euler	Sturm		
France	l'Hospital	Clairaut / Laplace / Lagrange / Legendre	Cauchy		
Poland			Wronski		

167

Index

A

advancing subscripts 116
arbitrary constant 3
auxiliary equation 77, 85

B

basis for the solution space 68, 73, 74
basis functions 68, 74
basis functions of the solution space 74
basis vectors 74
beer froth rate of disappearance 4
bell curve 25
Bernoulli brothers 37
Bernoulli, John 11
Bernoulli's equation 55
Bessel functions 121
Bessel's equation 118
boundary value 28
boundary value problem (BVP) 27, 28

C

candidate component functions 95
Cauchy, Augustin 113
Cauchy-Euler equation 109
chain rule 12, 37
circuit
      discrete-time 23
      LC 79
      LRC 80
      RL 47
Clairaut, Alexis 37
Clairaut's condition 38, 39
Clairaut's Theorem 38
Classifying equations and solutions 17
closed under differentiation 95

coefficient generator 120
compatible powers 116
complex conjugate roots 85
complex function 62
complex numbers 61
conjugate 62
conjugate of a complex function 62
constant function 38
continuity 33

D

degrees of freedom 21
derivative of step function 141
derivatives, Laplace transforms of 135-137
difference equation 22
differential equation 3
     linear 19, 20
     nonlinear 20
differential 3
     exact 41
differentiation, closed under 95
differentiation via multiplication 135
digital signal processing 22
dilation equations 23
Dirac delta function 151, 153
discontinuous functions 134
discontinuous square wave 134

E

Einstein, Albert 5, 8, 28, 29
Einstein's postulate 28, 29
Einstein's Special Theory of Relativity 8, 15, 28, 29
Einstein's Theory of Heat Capacity 5
elastic string fixed at both ends 75
electrical resonance 80
electromotive force 47
Euclidean two-space 67
Euler, Leonhard 61, 89
Euler's formula 61
exact differential 41
exact equation 37

impulse of unit area 152
impulse response 150
indefinite integral 3
independent basis vectors 68, 74
initial value 25
initial value problem (IVP) 25
inner product 66
inner product between functions 65
input domain 127
integrating factor 43
inverse first translation theorem 146
inverse-Laplace transform 125, 126
inverse transform 125

K

Kirchoff's Law 47, 80

L

LC circuit 80
LRC circuit 80
Lagrange, Joseph-Louis 103
Laplace domain 125
Laplace parameter 144
Laplace, Pierre-Simon 143
Laplace Transform 23, 125, 126
Laplace Transform translation theorems 143
Laplace transforms of derivatives 135-137
Leibnitz differential notation 3, 11
Leibnitz differentials 3, 11
Leibnitz, Gottfried 11, 47
Leibniz's rule 139
linear dependence 69
linear differential equation 19, 20
      n^{th} –order 70
linear differential equations of the first order (method of) 43
linear equation 43
linear independence 68
linear operator 115
linearity of the Laplace transform 130
linearly independent basis vectors 68, 69, 74, 85, 95
linearly independent solutions 74

Lorentz contraction 29
Lorentz Transformation 9

M

Maclaurin expansion 118

N

n-parameter family of solutions 21
neutron decay 8
Newton, Isaac 11
Newton's fly speck notation 11, 18
non-constant coefficients 101
non-homogeneous equation 76
non-homogeneous linear ODE with constant coefficients 95
non-trivial solution 90
normal form 19
norm of a complex number 62
norm of a function 66
notation
     dot 11, 18
     fly speck 11, 18
     primed 3

O

order of a differential equation 18
order of linearity 19
ordinary differential equations 17
orthogonal functions 65, 69
orthogonal vectors 69

P

partial differential equations 17
particular solution 22
Picard, Émile 34
Picard's Theorem 32
piecewise defined step 134
Poisson 126

R

radio 80
RL circuit 47
rational functions 132
ratios of polynomials 132
real world domain 5
recursion formula 117
reduction of order 82, 89
relativity equation 15
relativity, special 8, 15, 28
reverse transform 125

S

second translation theorem 154
separable variables 11, 12
series solution 115
solution 4, 7, 20
     existence of 31
     explicit 21
     general 70, 72
     homogeneous 78, 79, 95
     implicit 21
     non-trivial 90
     particular 78
     trivial 21
     uniqueness of 31, 71
solution space 67, 72, 73, 85
     basis for the 70
solving a differential equation 5, 7
space capsule 30
Special Relativity 8, 9, 15, 28
spring 4, 27
standard form 105
step function, derivative of 141
string
     vibrating 27, 28, 76
     elastic 75
struck harmonic oscillator 155
subscripts, advancing 116
superposition principle 73, 95
systems 3

174

T

U

V

W

www.ingramcontent.com/pod-product-compliance
Lightning Source LLC
Chambersburg PA
CBHW032011170526
45157CB00002B/641